PYTHON MACHINE LEARNING

Discover the Essentials of Machine Learning, Data Analysis, Data Science, Data Mining and Artificial Intelligence using Python Code with Python Tricks

Samuel Hack

Copyright © 2019 by Samuel Hack - All rights reserved

The book is only for personal use. No part of this publication may be reproduced, distributed, or transmitted in any form or by any means, including photocopying, recording, or other electronic or mechanical methods, without the prior written permission of the publisher, except in the case of brief quotations embodied in critical reviews and certain other noncommercial uses permitted by copyright law.

TABLE OF CONTENTS

INTRODUCTION 7

CHAPTER 1: FUNDAMENTALS OF MACHINE LEARNING 11

CHAPTER 2: MACHINE LEARNING ALGORITHMS 45

CHAPTER 3: BASICS OF DATA SCIENCE TECHNOLOGIES 87

CHAPTER 4: MACHINE LEARNING LIBRARY "SCIKIT-LEARN" 101 113

CHAPTER 5: NEURAL NETWORK TRAINING WITH TENSORFLOW 147

CHAPTER 6: DATA PRE-PROCESSING AND CREATION OF TRAINING DATA SET 173

CONCLUSION 211

INTRODUCTION

Congratulations on purchasing *Python Machine Learning: Discover the essentials of machine learning, data analysis, data science, data mining and artificial intelligence using Python Code with Python tricks* and thank you for doing so.

The following chapters will discuss the core concepts of "Machine Learning" models that are being developed and advanced using Python programming language. This book will provide you overarching guidance on how you can use Python to develop machine learning models using Scikit-Learn and TensorFlow machine learning libraries. You will start this book by gaining a solid understanding of the basics of machine learning technology and types of machine learning models. It is important to master the concepts of machine learning technology and learn how researchers are breaking the boundaries of data science to mimic human intelligence in machines using various learning algorithms. The power of machine learning technology has already started to manifest in our environment and in our everyday objects.

The chapter 2 titled "Machine Learning Algorithms", you will learn the nuances of "12 of the most popular machine learning algorithms", in a very easy to understand language that requires no background in Python coding language and indeed might spike your interest in this field of research. You will learn about the foundational machine learning algorithms namely, supervised, unsupervised and reinforcement machine learning algorithms that serve as the skeleton of hundreds of machine learning algorithms being developed every day.

The chapter 3, titled "Basics of Data Science technologies", will provide you a clear and concise overview of various Data Science technologies that are shaping our present and will dictate our technological future, as experiences through the Fourth Industrial revolution. Data science is an umbrella term used for the cutting-edge technologies of today such as "big data or big data analytics", "data mining technology", "machine learning technology" and even "artificial intelligence technology".

In chapter 4, titled "Machine Learning Library "Scikit-Learn" 101", we deep dive into the functioning of Scikit-Learn library along with the pre-requisites required to develop machine learning model using Scikit-Learn library. A detailed walkthrough with an open-source database using illustrations and actual Python code that you can try hands-on by following the instructions in this book. There is no better way to learn than to get your hands dirty and get real experience of the task. There is also guidance provided on resolving nonlinear issues with "k-nearest neighbor" and "kernel trick algorithms" in this book.

You will learn the entire process of creating of Neural Network models on TensorFlow machine learning platform, using open source data set for example, along with the actual Python code used for the development, in chapter 5, titled "Neural Network training with TensorFlow". Neural Networks are characterized by a "single neuron-like entity of the machine that is capable of learning the expected output for a given input from training data sets". TensorFlow is built on Python and touted as a simple and flexible architecture that supports the development of machine learning ideas from "concept to code to state-of-the-art models and publication" in a short time.

In the last chapter of the book, titled "Data pre-processing and Creation of training data set", you will learn all about the most time consuming and critical aspect of developing a machine learning model i.e. Data pre-processing and splitting the processed data set into training and testing subsets. Finally, as a bonus, you will learn some Python tips and tricks to take your machine learning programming game to the next level. So, breathe in, breathe out, and let's begin!

There are plenty of books on this subject on the market, thanks again for choosing this one! Every effort was made to ensure it is full of as much useful information as possible, please enjoy!

Chapter 1: Fundamentals of Machine Learning

The concept of Artificial Intelligence technology stems from the idea that machines are capable of human-like intelligence and can mimic human thought processing and learning capabilities, to adapt to new inputs and perform tasks without requiring human assistance. Machine learning is integral to the concept of artificial intelligence. Machine learning technology (ML) is referred to the concept of Artificial Intelligence technology that focuses primarily on the engineered capability of machines to explicitly learn and self-train, by identifying data patterns to improve upon the underlying algorithm and make independent decisions with no human intervention". In 1959, pioneering computer gaming and artificial intelligence expert, Arthur Samuel, coined the term "machine learning" during his tenure at IBM.

Machine learning hypothesizes that modern-day computers can be trained using targeted training data sets, that can be easily customized to develop desired functionalities. Machine learning is driven by the pattern recognition technique wherein the machine records and revisits past interactions and results, that are deemed in alignment with its current situation. Given the fact that machines are required to process the endless amount of data, with new data always pouring in, they must be equipped to adapt to the new data without needing to be programmed by a human, which speaks to the iterative aspect of ML.

Now the topic of machine learning is so "hot" that the academia, business world and the scientific community have their take on its definition. Here are some of the widely accepted definitions from select highly reputed sources:

- "Machine learning is the science of getting computers to act without being explicitly programmed." – Stanford University

- "The field of Machine Learning seeks to answer the question, how can we build computer systems that automatically improve with experience, and what are the fundamental laws that govern all learning processes?" – Carnegie Mellon University

- "Machine learning algorithms can figure out how to perform important tasks by generalizing from examples." – University of Washington

- "Machine Learning at its most basic is the practice of using algorithms to parse data, learn from it, and then decide or prediction about something in the world." – Nvidia

- "Machine learning is based on algorithms that can learn from data without relying on rules-based programming." – McKinsey.

BASIC CONCEPTS OF MACHINE LEARNING

The biggest draw of machine learning is the engineered capability of the system to learn programs from the data automatically instead of manually constructing the program for the machine. Over the last decade, the use of machine learning algorithms expanded from computer science to the industrial world. Machine learning algorithms are capable of generalizing tasks to execute them iteratively. The process of developing specific programs for specific tasks costs a lot of time and money, but occasionally it's just impossible to achieve. On the other hand, ML programming is often feasible and tends to be much more cost-effective. The use of machine learning in tackling ambitious issues of widespread importance such as global warming and depleting underground water levels is promising with a massive collection of relevant data.

"A breakthrough in machine learning would be worth ten Microsoft."

– Bill Gates

Many types of machine learning exist today but the concept of machine learning largely boils down to three components "representation", "evaluation" and "optimization". Here are some of the standard concepts that apply to all of them.

REPRESENTATION

Machine learning models are incapable of directly hearing, seeing or sensing input examples. Therefore, data representation is required to supply the model with a useful vantage point into the key qualities of the data. To be able to successfully train a machine learning model selection of key features that best represent the data is very important. "Representation" simply refers to the act of "representing" data points from a computer to a language that it understands using a set of classifiers. A classifier can be defined as "a system that inputs a vector of discrete and or continuous feature values and outputs a single discrete value called class". For a model to learn from the represented data the training data set or the "hypothesis space" must contain the desired classifier that you want the models to be trained on. Any classifiers that are external to the hypothesis space cannot be learned by the model.

The data features used to represent the input are very critical to the machine learning process. The data features are so important to the development of the desired machine learning model that can easily be the difference between successful and failed machine learning projects. A training data set containing multiple independent "features" that are well correlated with the "class" can make the machine learning much smoother. On the other hand, the class containing complex features may not be easy to learn from for the machine. This often requires the raw data to be processed so that desired features can be constructed from it, to be leveraged for the ML model. The process of deriving features from raw data tends to be the most time consuming and laborious part of the ML project. It is also considered the most creative and interesting part of the project where intuition and trial and error play just as important role as the technical requirements.

The process of ML is not a "one-shot process" of developing a training data set and executing it instead it is an iterative process that requires analysis of the post-run output, followed by modification of the training data set and then repeating the whole process all over again. Another reason for the extensive time and effort required to engineer the training data set is domain specificity. Training data set for an e-commerce platform to generate predictions based on consumer behavior analysis will be very different from the training data set required to develop a self-driving car. However, the actual machine learning process remains largely the same across industrial domains. No wonder, a lot of research is being done to automate the feature engineering process.

EVALUATION

Essentially the process of judging multiple hypothesis or models to choose one model over another is referred to as an evaluation. To be able to differentiate between useful classifiers from the vague ones an "evaluation function" is required. The evaluation function is also called as "objective", "utility" or "scoring" function. The machine learning algorithm has its internal evaluation function which tends to be different from the external evaluation function used by the researchers to optimize the classifier. Usually, the evaluation function is defined before the selection of the data representation tool, as the first step of the project. For example, the machine learning model for a self-driving car has the feature for identification of pedestrians in its vicinity at near-zero false negatives and a low false-positive as an evaluation function and the pre-existing condition that needs to be "represented" using applicable data features.

OPTIMIZATION

The process of searching the space of presented models to achieve better evaluations or highest-scoring classifier is called as "optimization". For algorithms with more than one optimum classifier, the selection of optimization technique is very critical in the determination of the classifier produced and to achieve a more efficient learning model. A variety of off-the-shelf optimizers are available in the market to kick start new machine learning models before eventually replacing them with custom-designed optimizers.

Table 1. The three components of learning algorithms.

Representation	Evaluation	Optimization
Instances	Accuracy/Error rate	Combinatorial optimization
K-nearest neighbor	Precision and recall	Greedy search
Support vector machines	Squared error	Beam search
Hyperplanes	Likelihood	Branch-and-bound
Naive Bayes	Posterior probability	Continuous optimization
Logistic regression	Information gain	Unconstrained
Decision trees	K-L divergence	Gradient descent
Sets of rules	Cost/Utility	Conjugate gradient
Propositional rules	Margin	Quasi-Newton methods
Logic programs		Constrained
Neural networks		Linear programming
Graphical models		Quadratic programming
Bayesian networks		
Conditional random fields		

BASIC MACHINE LEARNING TERMINOLOGIES

Agent – In the context of reinforcement learning, an agent refers to "the entity that uses a policy to maximize expected return gained from transitioning between states of the environment".

Boosting – Boosting can be defined as "a machine learning technique that iteratively combines a set of simple and not very accurate classifiers (referred to as weak classifiers) into a classifier with high accuracy (a strong classifier) by up-weighting the examples that the model is currently misclassifying".

Candidate generation – The phase of selecting the "first set of recommendations" by a recommendation system is referred to as candidate generation. For example, a book library can offer 500,000 titles. This technique will produce a subset of few 100 books meeting the needs of a particular user and can be refined further to an even smaller set as needed.

Categorical Data – Data features boasting a "discrete set of possible values" is called as categorical data. For example, a "categorical feature" labeled car style can have an unconnected set of multiple possible values including sedan, coupe, SUV.

Checkpoint – Checkpoint can be defined as "The data that can capture the state of the variables of a learning model particular moment in time". With the use of checkpoints, training can be carried out across multiple sessions and model weights or scores can be exported.

Class – Class can be defined as "one of a set of listed target values for a given label". For example, a machine learning model designed to detect "spam" will have two classes, namely, "spam" and "not spam".

Classification model – The type of ML model used to "distinguish between two or more discrete classes of data" is referred to as a classification model. For example, a classification model for identification of dog breeds could assess whether the dog picture used as input is Labrador, Schnauzer, German Shepherd, Beagle and so on.

Collaborative filtering – The process of generating predictions for a particular user based on the shared interests of a group of similar users is called collaborative filtering.

Continuous feature – It is defined as a "floating-point feature with an infinite range of possible values".

Discrete feature – It is defined as a "rigid feature with a finite set of possible values".

Discriminator – "The system that determines whether the input examples are real or fake" is called discriminator.

Down-sampling – The process of Down-sampling refers to "the act of reducing the amount of information contained in a feature or using a disproportionately low percentage of over-represented class examples to train the learning model more efficiently".

Dynamic model – A learning model that is continuously receiving input data to be trained continuously is called dynamic model.

Ensemble – "The set of predictions generated by merging predictions of multiple models" is called ensemble.

Environment – The term 'environment' used in the context of reinforcement machine learning constitutes "the world that contains the agent and allows the agent to observe that world's state".

Episode – The term episode used in the context of reinforcement machine learning constitutes "every iterative attempt made by the agent to learn from its environment".

Feature – "An input data variable that is used in generating predictions" is called feature.

Feature engineering – Feature engineering can be defined as "the process of determining which features might be useful in training a model, and then converting raw data from log files and other sources into said features".

Feature extraction – Feature extraction can be defined as "the process of Retrieving intermediate feature representations calculated by an unsupervised or pre-trained model for use in another model as input".

Few-shot learning - Few-shot learning can be defined as "a machine learning approach, often used for object classification, designed to learn effective classifiers from only a small number of training examples".

Fine-tuning – The process of "performing a secondary optimization to adjust the parameters of an already trained model to fit a new problem" is called as fine-tuning.

It is widely used to refit the weights of a "trained unsupervised model" to a "supervised model".

Generalization – "The ability of the machine learning model to make correct predictions on new, previously unseen data as opposed to the data used to train the model" is called generalization.

Inference – In the context of ML, an inference can be defined as "the process of making predictions by applying the trained model to unlabeled examples".

Label – In the context of machine learning (supervised), the "answer" or "result" part of an example is called label. A labeled data set can constitute single or multiple features and corresponding labels for every example. For example, in a house data set, the features could include the year built, number of bedrooms and bathrooms, while the label can be the "house's price".

Linear model – Linear model is defined as "a model that assigns one weight per feature to make predictions".

Loss – In the context of machine learning, loss refers to the "measure of how far are the predictions generated by the model from its label".

Matplotlib – It is "an open-source Python 2D plotting library which is used to visualize different aspects of machine learning".

Model – In the context of ML, a model can be defined as "the representation of what a machine learning system has learned from the training data".

NumPy – "An open-source math library that provides efficient array operations in Python".

One-shot learning – In the context of ML, one-shot learning refers to "a machine learning approach designed to learn effective classifiers from a single training example, often used for object classification".

Overfitting - In the context of machine learning, overfitting is referred to as "creation of a model that matches the training data so closely that the model fails to make correct predictions on new data".

Parameter – "A variable of a model that the machine learning system can train on its own" is called parameter.

Pipeline – In the context of ML, pipeline refers to "the infrastructure surrounding a machine learning algorithm and includes a collection of data, addition of the data to training data files, training one or more models, and exporting the models to production".

Random forest – In the context of machine learning, the concept of random forest pertains to "an ensemble approach for finding the decision tree that best fits the training data by developing multiple decision trees with a random selection of features".

Scaling - In the context of machine learning, scaling refers to "a common feature engineering practice to tame a feature's range of values to match the range of other features in the dataset".

Sequence model - A sequence model simply refers to a model with sequential dependency on data inputs to generate a future prediction.

Under-fitting - In the context of ML, under-fitting refers to "production of a ML model with poor predictive ability because the model hasn't captured the complexity of the training data".

Validation – "The process used to evaluate the quality of a machine learning model using the validation set, as part of the model training phase" is called as validation. The main goal of this process is to make sure that the performance of the ML model can be applied beyond the training set.

TYPES OF MACHINE LEARNING MODELS

Artificial Neural Networks

"Artificial Neural Networks" or (ANN) have been developed with inspiration from the structure of the human brain and utilizes numerous processing units like human neurons or nodes working as one. Each ANN has numerous concealed layers including an input and an output layer. These neurons are connected in a complex with one another by weighted links. A solitary neuron is capable of accepting input from various neurons. When a neuron gets activated, it casts a "weighted vote" to control the activation of the subsequent neuron that is collecting that input. An algorithm modifies these weights following the training data to optimize learning. A simple algorithm called "fire together, wire together" increases the weight between two connected neurons when the activation of either neuron leads to subsequent activation of the other neuron. "Concepts" are formed and distributed through the sub-network of shared neurons.

The most widely recognized ANN deal with the unidirectional progression of information and are called "Feed forward ANN". However, ANN can also be used for bidirectional and cyclic progression of information to gain state equilibrium. ANNs are less dependent on prior assumptions and able to learn from prior cases by modifying the connected weights. They can be utilized with either supervised or unsupervised learning. The use of supervised learning will generate correct ANN output for every input pattern. By varying the weights, the error between the target output and the output produced by ANN can be effectively minimized. For example, a type of "supervised learning algorithm" called "reinforced learning", informs the ANN whether the output it produced is correct or not instead of directly supplying the correct output. On the other hand, "unsupervised learning algorithm" is capable of providing a variety of input pattern to the ANN which is then processed by the ANN to self-explore the relationship between the provided input patterns and learn to categorize them accordingly. Artificial Neural Networks utilizing a combination of "supervised" and "unsupervised" learning are also available.

ANN is the choice of the learning model to address data-driven problems with unknown or difficult to comprehend algorithm or rules, credited to their defined data structure and non-linear computations. ANNs are capable of efficiently processing complex information and can easily withstand multi-variable data errors. Problems requiring a complete understanding of and insight into the actual process are not suitable for the ANN model, given its black box nature providing no view into the underlying process.

ANNs are widely used to resolve problems that require:

- "Pattern classification" by assigning input pattern to one of the pre-determined classes. For example, classification of land-based on satellite images.

- "Clustering", which is simply an unsupervised pattern classification model. For example, the prediction of the ecological status of water streams using defined input patterns.

- "Function approximation or regression", which is capable of creating a function out of the provided set of training patterns. For example, prediction of ozone concentration in the atmosphere, estimation of the amount of nitrate in groundwater and modeling water supply.

- "Prediction", which utilized prior data samples in a time series to estimate the output. For example, prediction of air and water quality.

- "Optimization", which is used to maximize or minimize a "cost function" subject to predefined constraints. For example, calibration of infiltration equations.

- "Retrieval by content", which is capable of recalling memory even if the input is incomplete or distorted. For example, using satellite images to produce water quality proxies.

- "Process control", which seeks to keep the velocity under changing data load close to constant by changing the throttle angle. For example, engine speed control.

Genetic Algorithms

As one can gather from the name, Genetic Algorithm (GA) mimics the nature's theory of selection by transferring the traits of the fitter solutions to the "offspring" and supplanting the less fit solutions. The Genetic Algorithm will keep evolving until a desired solution of the problem is achieved. Similar to human chromosomes, every potential solution is encoded as a "binary string" of traits and each arrangement of the progressive population are referred to as "generations". The original population set is produced randomly and all superseding generations are created through the process of selection and reproduction. A specific subset of the populace is then selectively bred to produce new chromosomes. The process of selection is driven by the fitness of the individual solutions including closeness to a perfect solution and deterministic examining. The closeness to a perfect selection is frequently carried out by the "roulette selection" method leading to the arbitrary selection of a parent with calculated probability based on its fitness. Then the deterministic examining allots a value to a subset of a selected organism.

Conventional propagation is achieved through genetic crossover, which produces the off-spring by trading chromosomes from two parents and inserting mutations in the chromosome to randomly modify some portion of the parent chromosome. The propagation doesn't happen as often as in the human world but allows the introduction of new genetic material in the gene pool. Mutation is considered less important than crossover in the advancement of the search but is deemed vital in the maintenance of genetic diversity, which is fundamental to continued evolution. In steady-state genetic algorithms, fewer fit members are superseded by the new generation, bringing about an increase in the average fitness of the model. The cycle of "reproduction and selection" is repeated until the completion criteria are met, for example, optimum fitness has been reached or all organisms are identical and evolution is not generating new results.

By focusing only on fitness examination and not accounting for other derivatives, genetic algorithms are considered highly robust computationally and capable of can smoothly balance load and efficacy. Another significant aspect of the genetic algorithm is its capability to indirectly sample a big volume of code sequences that have been tested. Unlike the stochastic search techniques which cannot search through noisy and multimodal relations, genetic algorithm can store a whole population of solutions instead of adjusting just a single solution. Some application examples of genetic algorithms include: "forecasting air quality, calibration of the water-quality models, estimation of soil bulk density and water management systems".

Decision trees - A machine learning decision tree can be defined as "a tree-like graphical representation of the decision-making process, by taking into consideration all the conditions or factors that can influence the decision and the consequences of those decisions". Decision trees are considered one of the simplest "supervised machine learning algorithms" and has three main elements: "branch nodes" representing conditions of the data set, "edges" representing ongoing decision process and "leaf nodes" representing the end of the decision. Decision trees make for an excellent predictive analysis technique with wide application in generating predictions for the values of "categorical" as well as "continuous" target variables. (Note – The very important concept of Decision Trees will be explored in details in the next chapter of this book)

Probabilistic Programming

"Probabilistic programming" is a high-level programming language that empowers formation of probability models, with the capability to extract values from distributions and condition these values into a program. The Probabilistic programming based learning systems are capable of making inferences from earlier knowledge, permitting decision making even in the face of uncertainty. To capture the knowledge of the target system "quantitative" and "probabilistic" terms are utilized. With adequate training, the model can be applied to explicit inquires to generate answers through a process called "inference".

With the use of probabilistic programming language, the probability model can also be solved automatically without any external assistance. It also supports uninterrupted access and reusability of machine learning model libraries and provides support for interactive modeling as well as formal verification. The abstraction layer within the probabilistic algorithm is paramount to nurture generic and efficient inferences because various Artificial Intelligence problems require the agent to process fragmented and distorted data set. Probabilistic algorithms are widely used to analyze large volumes of data to generate predictions and assist perception systems in the analysis of the underlying processes. Some examples of Probabilistic programming applications are: "medical imaging", "financial predictions", "machine perception" and "weather forecasting".

The most widely used probabilistic model is called "Bayesian network", which uses "Bayesian inference" to compute probability. The Bayesian network can generate inferences in two forms. The first one is the assessment of the joint probability of a specific assignment of values for every single variable in the network. While the second form is the assessment of the probability of a subset of variables that have been given assignments of an alternate subset of variables. The problems about reasoning, learning, planning, and perception of the system can be addressed with the application of the Bayesian networks. The "Bayesian inference algorithm" is utilized in the resolution of reasoning problems while the "expectation" and "maximization" algorithm can be used to address learning problems. The dynamic Bayesian networks allow for unfettered perception and decision networks which are widely used to solve planning issues.

MACHINE LEARNING IN PRACTICE

The complete process of machine learning is much more extensive than just the development and application of machine learning algorithms and can be divided into steps below:

1. Define the goals of the project taking into careful consideration all the prior knowledge and domain expertise available. Goals can easily become ambiguous since there are always additional things you want to achieve than practically possible to implement.

2. The data pre-processing and cleaning must result in a high-quality data set. This is the most critical and time-consuming step of the whole project. The larger the volume of data, the more noise it brings to the training data set which must be eradicated before feeding to the learner system.

3. Selection of appropriate learning model to meet the requirements of your project. This process tends to be rather simple given the various types of data models available in the market.

4. Depending on the domain the machine learning model is applied to, the results may or may not require a clear understanding of the model by human experts as long as the model can successfully deliver desired results.

5. The final step is to consolidate and deploy the knowledge or information gathered from the model to be used on an industrial level.

6. The whole cycle from step 1 to 5 listed above is iteratively repeated until a result that can be used in practice is achieved.

IMPORTANCE OF MACHINE LEARNING

To get a sense of how significant machine learning is in our everyday lives, it is simpler to state what part of our cutting edge way of life has not been touched by it. Each aspect of human life is being impacted by the "smart machines" intended to expand human capacities and improve efficiencies. Artificial Intelligence and machine learning technology is the focal precept of the "Fourth Industrial Revolution", that could question our thoughts regarding being "human".

Here are a few reasons to help you understand the significance of machine learning in our daily lives:

- Automation of repetitive learning and revelation from data. Not at all like hardware-driven robotic automation that simply automates manual assignments, machine learning allows performance of high volume, computer-based tasks consistently and dependably.

- Machine learning algorithms are helping Artificial Intelligence to adapt to the evolving world by allowing the machine or system to learn, take note and improve upon its prior errors. Machine learning algorithm functions as a classifier or a predictor to acquire new skills and identify data pattern and structure. For example, machine learning algorithm has generated a system that can teach itself how to play chess and even how to generate product recommendations based on customer activity and behavior data. The beauty of this model is that it adapts with every new set of data.

- Machine learning has analyzed deeper and larger data set feasible with the use of neural networks containing multiple hidden layers. Think about it, a fraud detection system with numerous concealed layers would deem a work of fantasy just a couple of years ago. With the advent of big data and unlikely to envision computer powers, a whole new world is on the horizon. The data of the machines resemble the gas to the vehicle, the more data you can add to them, faster and more accurate results will get. Deep learning models flourish with an abundance of data because they gain straightforwardly from the data.

- The "deep neural networks" of the machine learning algorithms have resulted in unbelievable accuracy. For example, frequent and repeated use of smart tech like "Amazon Alexa" and "Google Search", results in increased accuracy derived from deep learning. These "deep neural networks" are also empowering our medical field. Image classification and object recognition are now capable of finding cancer on MRIs with similar accuracy as that of a highly trained radiologist.

- Artificial Intelligence is allowing for enhanced and improved use of big data analytics in conjunction with machine learning algorithms. Data has evolved as its currency and when algorithms are self-learning it can easily become "intellectual property". The raw data is similar to a gold mine in that the more and deeper you dig, the more "gold" or valuable insight you can dig out or extract. Application of machine learning algorithms to the data can enable you to find the correct solutions quicker and makes for an upper hand. Keep in mind the best information will consistently win, although everyone is utilizing comparative techniques.

Chapter 2: Machine Learning Algorithms

By utilizing prior computations and underlying algorithms, machines are now capable of learning from and training on their own to generate high-quality, readily reproducible decisions and results. The notion of machine learning has been around for a long time now, but the latest advances in machine learning algorithms have made large data processing and analysis feasible for computers. This is achieved by applying sophisticated and complicated mathematical calculations using high speed and frequency automation. Today's advanced computing machines can analyze humongous information quantities quickly and deliver quicker and more precise outcomes. Companies using machine learning algorithms have increased flexibility to change the training data set to satisfy their company needs and train the machines accordingly. These tailored algorithms of machine learning enable companies to define potential hazards and possibilities for development. Typically, machine learning algorithms are used in cooperation with artificial intelligence technology and cognitive techniques to create computers extremely efficient and extremely effective in processing large quantities of information or big data and to generate extremely precise outcomes.

There are four fundamental types of machine learning algorithms available today.

SUPERVISED MACHINE LEARNING ALGORITHMS

Due to their ability to evaluate and apply the lessons learned from prior iterations and interactions to fresh input data set, the supervised learning algorithms are commonly used in predictive big data analysis. Based on the instructions given to effectively predict and forecast future occurrences, these algorithms can label all their ongoing runs. For instance, people can program the machine as "R" (Run), "N" (Negative) or "P" (Positive) to label its data points. The algorithm for machine learning will then label the input data as programmed and obtain data inputs with the right outputs. The algorithm will compare its own produced output to the "anticipated or correct" output, identifying future changes that can be created and fixing mistakes to make the model more precise and smarter. By using methods such as "regression," "prediction," "classification" and "ingredient boosting" to train the machine learning algorithms well, any new input data can be fed into the machine as a set of "target" data to orchestrate the learning program as desired. This "known training data set" jump-starts the analytical process followed by the learning algorithm to produce an "inferred feature" that can be used to generate forecasts and predictions based on output values for future occurrences. For instance, financial institutions and banks rely strongly on monitoring machine learning algorithms to detect credit card fraud and predict the probability of a prospective credit card client failing to make their credit payments on time.

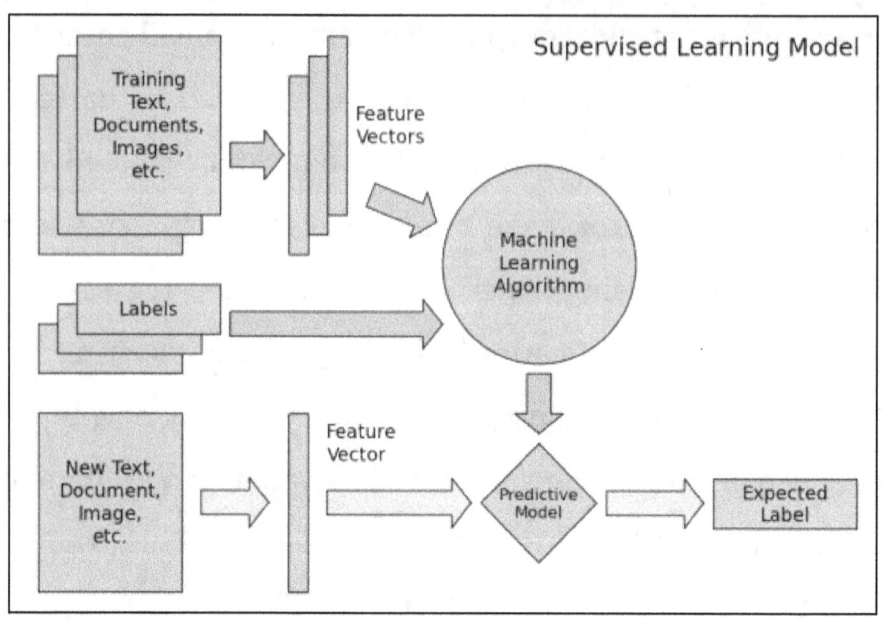

UNSUPERVISED MACHINE LEARNING ALGORITHMS

Companies often find themselves in a scenario where information sources are inaccessible that are needed to produce a labeled and classified training data set. Using unsupervised ML algorithms is perfect in these circumstances. Unsupervised ML algorithms are widely used to define how the machine can generate "inferred features" to elucidate a concealed construct from the stack of unlabeled and unclassified data collection. These algorithms can explore the data to define a structure within the data mass. Unlike the supervised machine learning algorithms, the unsupervised algorithms fail to identify the correct output, although they are just as effective as the supervised learning algorithms in investigating input data and drawing inferences. These algorithms can be used to identify information outliers, generate tailored and custom product recommendations, classify text subjects using methods such as "self-organizing maps," "singular value decomposition" and "k-means clustering." For instance, customer identification with shared shopping characteristics that can be segmented into particular groups and focused on comparable marketing strategies and campaigns. As a result, in the online marketing world, unsupervised learning algorithms are extremely popular.

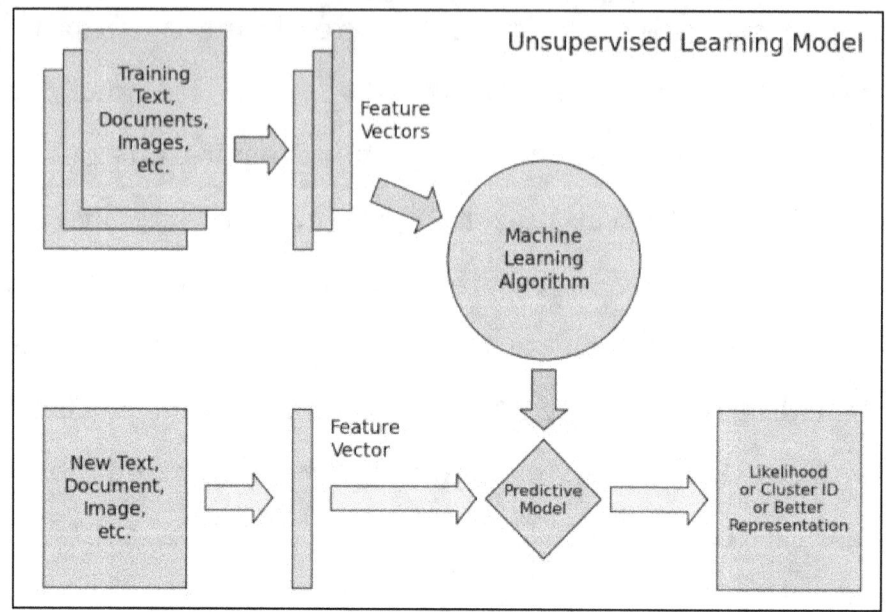

SEMI-SUPERVISED MACHINE LEARNING ALGORITHMS

Highly versatile, the "semi-supervised machine learning algorithms" are capable of using both labeled and unlabeled information set to learn from and train themselves. These algorithms are a "hybrid" of algorithms that are supervised and unsupervised. Typically, with a small volume of labeled data, the training data set is comprised of predominantly unlabeled data. The use of analytical methods including "forecast," "regression" and "classification" in conjunction with semi-supervised learning algorithms enable the machine to considerably enhance its learning precision and training capabilities. These algorithms are commonly used in instances where it is highly resource-intensive and less cost-effective for the business to generate labeled training data set from raw unlabeled data. Companies use semi-supervised learning algorithms on their systems to avoid incurring extra costs of staff and equipment. For instance, the implementation of "facial recognition" technology needs a huge amount of facial data distributed across various sources of input. The raw data pre-processing, processing, classification and labeling, acquired from sources such as internet cameras, needs a lot of resources and hours of work to be used as a training data set.

REINFORCEMENT MACHINE LEARNING ALGORITHMS

The "reinforcement machine learning algorithms" are much more distinctive in that they learn from the environment. These algorithms conduct activities and record the outcomes of each action diligently, which would have been either a failure resulting in mistake or reward for good performance. The two primary features that differentiate the reinforcement learning algorithms are the research method of "trial and error" and feedback loop of "delayed reward." Using a range of calculations, the computer constantly analyzes input data and sends a reinforcement signal for each right or anticipated output to ultimately optimize the final result. The algorithm develops a straightforward action gaining feedback loop to evaluate, record and learn which actions have been effective and in a shorter time have resulted in correct or expected output. The use of these algorithms allows the system to automatically determine optimal behaviors and maximize its efficiency within the constraints of a particular context. The reinforcement machine learning algorithms are therefore strongly used in gaming, robotics engineering, and navigation systems.

The machine learning algorithms have proliferated to hundreds and thousands and counting. Here are some of the most widely used algorithms:

1. Regression

The methods of "regression" fall under the supervised machine learning category. They assist in predicting or describing a specific numeric value based on the set of preceding data, such as anticipating a property's price based on preceding price data for comparable properties. Regression methods range from simple (such as "linear regression") to complex (such as "regular linear regression", "polynomial regression", "decision trees", "random forest regression" and "neural networks", among others).

The easiest technique of all is "linear regression", where the "mathematical equation of the line (y= m* x+ b) is used to model a data set". A "linear regression" model can be trained with multiple "data pairs (x, y)" by calculating a line's position and slope that can reduce the overall distance between data points and the line. In other words, for a line that generates the best approximation for the data observations, the calculation of the "slope (m)" and "y-intercept (b)" is used.

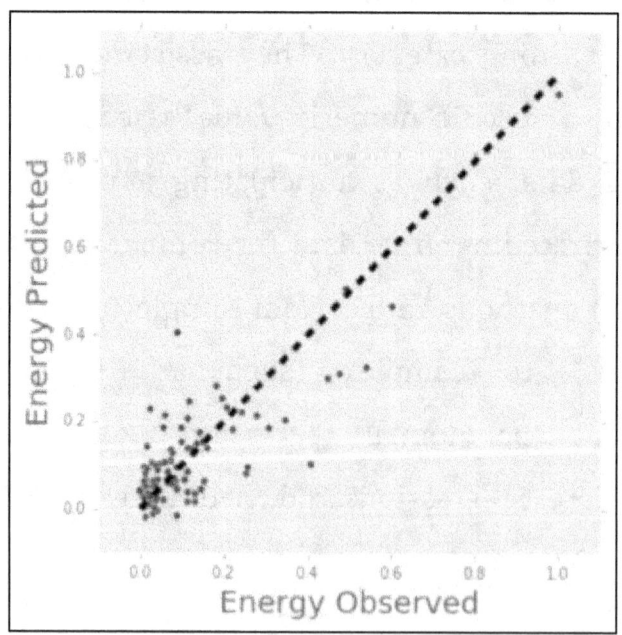

For example, using "linear regression" technique to generate predictions for the energy consumption (in kWh) of houses by collecting the age of the house, some bedrooms, square footage area, and several installed electronic equipment. Now, we have more than one input (year built, square footage) it is possible to use "linear multi-variable regression". The underlying process is the same as "one-to-one linear regression", however, the "line" created was based on the number of variables in multi-dimensional space.

The above plot demonstrates how well this model of linear regression fits the real construction energy consumption. In case you could gather house characteristics such as year built and square footage, but you don't understand the house's energy consumption then you are better off using the fitted line to generate approximations for the house's energy consumption.

2. Classification

The method of "classification" is another class of "supervised machine learning", which can generate predictions or explanations for a "class value". For example, this method can be used to predict the if of an online customer will purchase a particular product. The result generated will be reported as a yes or no response i.e. "buyer" or "not a buyer". But techniques of classification are not restricted to two classes. A classification technique, for instance, could assist to evaluate whether a specified picture includes a sedan or a SUV. The output will be three different values in this case: 1) the picture contains a sedan, 2) the picture contains a SUV, or 3) the picture does not contain either a sedan or a SUV.

"Logistic regression" is considered the easiest classification algorithm, though the term comes across as a "regression" technique that is far from reality. "Logistic regression" generates estimations for the likelihood of an event taking place based on single or multiple input values. For example, to generate estimation for the likelihood of a student being accepted to a specific university, a "logistic regression" will use the standardized testing scores and university testing score for a student as inputs. The generated prediction is a probability, ranging between '0' and '1', where 1 is full assurance. For the student, if the estimated likelihood is greater than 0.5, then the prediction would be that they will be accepted. If the projected probability is less than 0.5, the prediction would be that they will be denied admission.

The following graph shows the ratings of past learners as well as whether they have been accepted. Logistic regression enables creation of a line that can represent the "decision boundary".

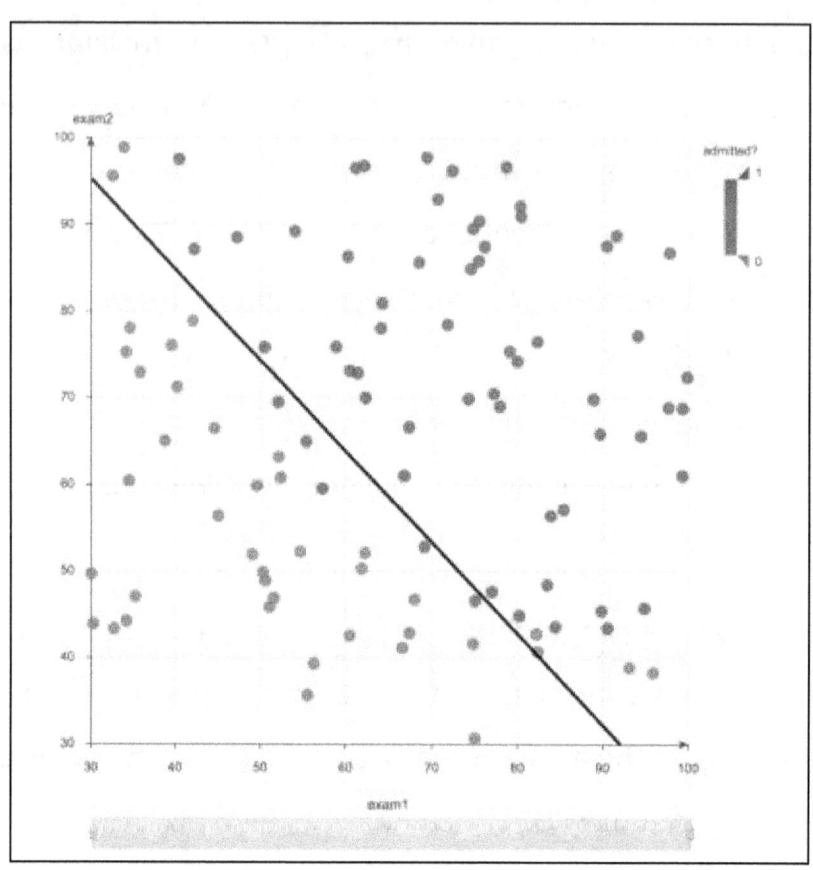

3. Clustering

We enter the category of unsupervised machine learning, with "clustering methods" because its objective is to "group or cluster observations with comparable features". Clustering methods do not use output data to train but allow the output to be defined by the algorithm. Only data visualizations can be used in clustering techniques to check the solution's quality.

"K-Means clustering", where 'K' is used to represent the number of "clusters" that the customer elects to generate and is the most common clustering method. (Note that different methods for selecting K value, such as the "elbow technique", are available.)

Steps used by K-Means clustering to process the data points:

1. The data centers are selected randomly by 'K'.

2. Assigns each data point to the nearest centers that have been randomly generated.

3. Re-calculates each cluster's center.

4. If centers do not change (or have minor change), the process will be completed.

Otherwise, we'll go back to step 2. (Set a maximum amount of iterations in advance to avoid getting stuck in an infinite loop, if the center of the cluster continues to alter.)

The following plot applies "K-Means" to a building data set. Each column in the plot shows each building's efficiency. The four measurements relate to air conditioning, heating, installed electronic appliances (refrigerators, TV) and cooking gas. For simplicity of interpretation of the results, 'K' can be set to value '2' for clustering, wherein one cluster will be selected as an efficient building group and the other cluster as an inefficient building group. You see the place of the structures on the left as well as a couple of the building characteristics used as inputs on the right: installed electronic appliances and heating.

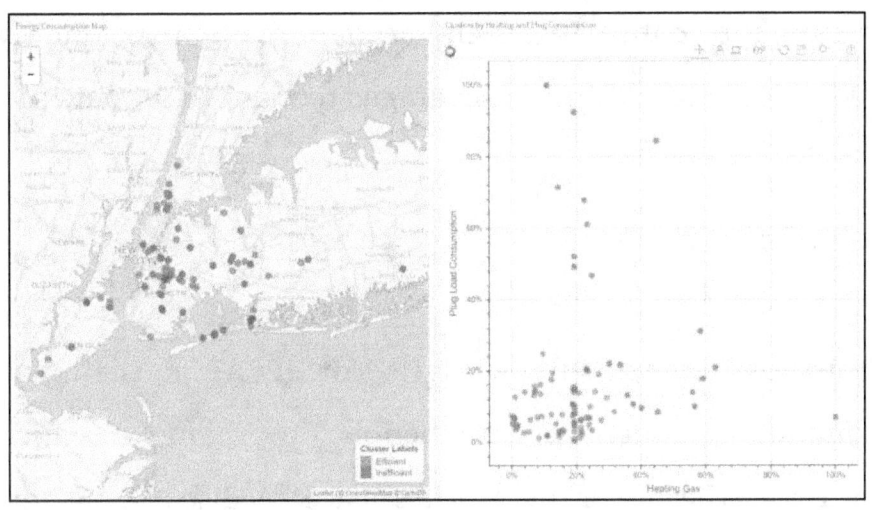

4. Dimension Reduction

As the name indicates, to extract the least significant information (sometimes redundant columns) from a data set, we use "dimensionality reduction". In practice, data sets tend to contain hundreds or even thousands of rows (also known as characteristics), which makes it essential to reduce the total number of rows. For example, pictures may contain thousands of pixels, all those pixels are not important for the analysis. Or a large number of measurements or experiments can be applied to every single chip while testing microchips within the manufacturing process, the majority of which produce redundant data. In such scenarios, "dimensionality reduction" algorithms are leveraged to manage the data set.

Principal Component Analysis

"Principal Component Analysis" or (PCA) is the most common "dimension reduction technique", which decreases the size of the "feature space" by discovering new vectors that are capable of maximizing the linear variety of the data. When the linear correlations of the data are powerful, PCA can dramatically decrease the data dimension without losing too much information. PCA is one of the fundamental algorithms of machine learning. It enables you to decrease the data dimension, losing as little information as possible. It is used in many fields such as object recognition, vision of computers, compression of information, etc. The calculation of the main parts is limited to the calculation of the initial data's vectors and covariance matrix values or of the data matrix's unique decomposition. Through one we can convey several indications, merge, so to speak, and operate with a simpler model already. Of course, most probably, data loss will not be avoided, but the PCA technique will assist us to minimize any losses.

t-Stochastic Neighbor Embedding (t-SNE)

Another common technique is "t-Stochastic Neighbor Embedding (t-SNE)", which results in a decrease of non-linear dimensionality. This technique is primarily used for data visualization, with potential use for machine learning functions such as space reduction and clustering.

The next plot demonstrates "MNIST database" analysis of handwritten digits. "MNIST" includes a large number of digit pictures from 0 to 9, used by scientists to test "clustering" and "classification" algorithms. Individual row of the data set represents "vectorized version" of the original picture (size 28x28 = 784 pixels) and a label (0, 1, 2 and so on) for each picture. Note that the dimensionality is therefore reduced from 784 pixels to 2-D in the plot below. Two-dimensional projecting enables visualization of the initial high-dimensional data set.

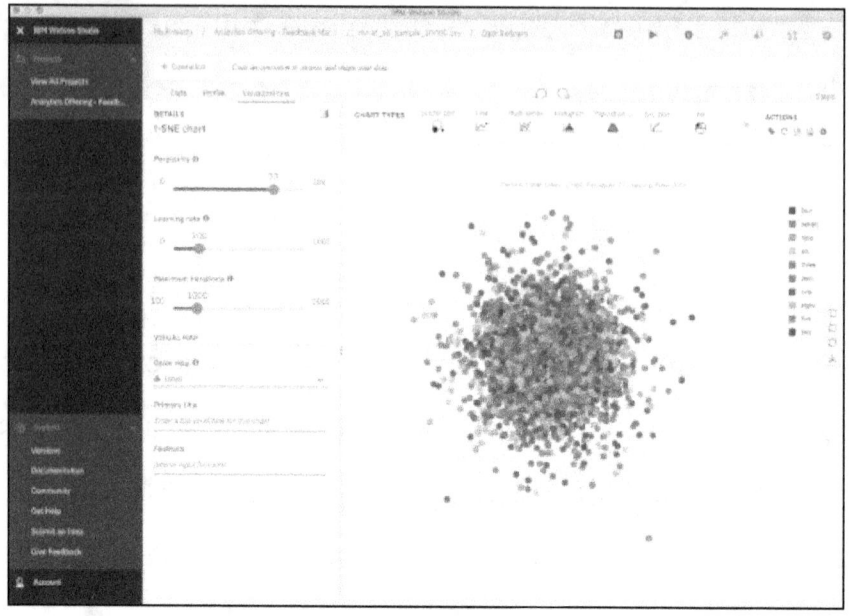

5. Ensemble Methods

Think that you have chosen to construct a car because you are not pleased with the variety of cars available in the market and online. You may start by discovering the best option for every component that you need. The resulting car will outshine all the other alternatives with the assembly of all these excellent components.

"Ensemble methods" use the same concept of mixing several predictive models (controlled machine learning) to obtain results of greater quality than any of the models can generate on their own. The "Random Forest" algorithms, for instance, is an "ensemble method" that collates various trained "Decision Trees" with different data set samples. Consequently, the quality of predictions generated by "Random Forest" method is better than the quality of the estimated predictions generated by only one "Decision Tree".

Think of "ensemble methods" as an approach for reducing a single machine learning model's variance and bias. This is essential because, under certain circumstances, any specified model may be accurate but completely incorrect under other circumstances. The relative accuracy could be overturned with another model. The quality of the predictions is balanced by merging the two models.

6. Transfer Learning

Imagine you are a data scientist focusing on the clothing industry. You have been training a high-quality learning model for months to be able to classify pictures of "women's tops" as tops, tank tops, and blouses. You have been tasked to create a comparable model for classification of pants pictures such as jeans, trousers, and chinos. With the use of "Transfer Learning" method, the understanding incorporated into the first model be seamlessly transferred and applied to the second model.

Transfer Learning pertains to the re-use and adaptation of a portion of a previously trained neural network to a fresh but comparable assignment. Specifically, once a neural network has been successfully trained for a particular task, a proportion of the trained layers can be easily transferred and combined with new layers which are then trained on pertinent data for the new task. This new "neural network" can learn and adapt rapidly to the new assignment by incorporating a few layers.

The primary benefit of transferring learning is decreasing in the volume of data required to train the neural network resulting in cost savings for development of "deep learning algorithms". Not to forget how hard it can be to even procure a sufficient amount of labeled data to train the model.

Suppose in this example, you are using a neural network with 20 hidden layers for the "women's top" model. You understand after running a few tests that 16 of the women's top model layers can be transferred and combined them with a new set of data to train on pants pictures. Therefore, the new pants model will have 17 concealed layers. The input and output of both the tasks are distinct, but the reusable layers are capable of summarizing the data appropriate to both, e.g. clothing, zippers, and shape of the garment.

Transfer learning is getting increasingly popular, so much so that for basic "deep learning tasks" such as picture and text classification, a variety of high quality pre-trained models are already available in the market.

7. Natural Language Processing

A majority of the knowledge and information of our world is in some type of human language. Once deemed as impossible to achieve, today computers are capable of reading large volumes of books and blogs within minutes. Although, computers are still unable to fully comprehend "human text", but they can be trained to perform specific tasks. Mobile devices, for instance, can be trained to auto-complete text messages or fix spelling mistakes. Machines have been trained enough to hold straightforward conversations like humans.

"Natural Language Processing" (NLP) is not exactly a method of ML, instead it is a commonly used technique to produce texts for machine learning. Consider a multitude of formats of tons of text files (words, internet blogs, etc). Most of these text files are usually flooded with typing errors, grammatically incorrect characters and phrases that need to be filtered out. The most popular text processing model available in the market today is "NLTK

(Natural Language ToolKit)", developed by "Stanford University" researchers.

The easiest approach to map texts into numerical representations concern calculation of the frequency of each word contained in every text document. For example, an integer matrix where individual rows represent one text document and every column a single word. This word frequency representation matrix is frequently referred to as the "Term Frequency Matrix" (TFM). From there, individual matrix entries can be separated by weight of how essential every single term is within the whole stack of papers. This form of matrix representation of a text document is called "Term Frequency Inverse Document Frequency" (TFIDF), which usually yields better performance for machine learning tasks.

8. Word Embedding

"Term Frequency Matrix" and "Term Frequency Inverse Document Frequency" are numerical representations of text papers which only take into account frequency and weighted frequencies to represent text files. On the other hand, "Word Embedding" in a document is capable of capturing the actual context of a word. Embedding can quantify the similarity between phrases within the context of the word, which subsequently allows the execution of arithmetic operations with words.

"Word2Vec" is a neural network-based technique that can map phrases to a numerical vector in a corpus. These vectors are then used to discover synonyms, do arithmetic with words or phrases, or to represent text files. Let's suppose, for instance, a large enough body of text files was used to estimate word embedding. Suppose the words "king, queen, man, and female" are found in the corpus and vector ("word") is the number vector representing the word "word". We can conduct an arithmetic procedure with numbers to estimate vector('woman'):

vector('king') + vector('woman') — vector('man') ~ vector('queen')

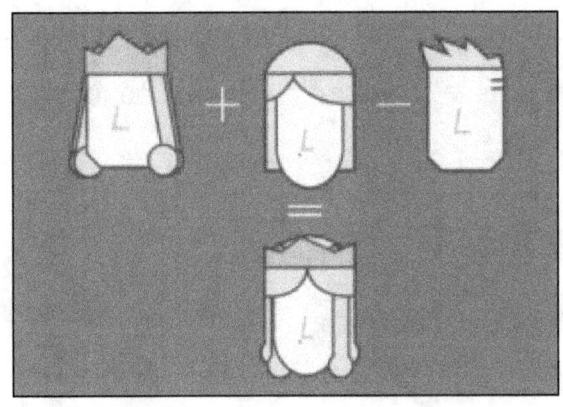

Word depictions enable similarities to be found between phrases by calculating the "cosine similarity" between the vector representation of the two words. The "cosine similarity" gives a measure of the angle between two vectors.

We use machine learning techniques to calculate word embedding, but this is often a preliminary step in implementing a machine learning algorithm on top of the word embedding method. For example, the "Twitter" user database containing a large volume of "tweets" can be leveraged to understand which of these customers purchased a house recently. We can merge "Word2Vec" with logistic regression to generate predictions on the likelihood of a new "Twitter" user purchasing a home.

9. Decision Trees

To refresh your memory; a machine learning decision tree can be defined as "a tree-like graphical representation of the decision-making process, by taking into consideration all the conditions or factors that can influence the decision and the consequences of those decisions". Decision trees are considered one of the simplest "supervised machine learning algorithms", with three main elements: "branch nodes" representing conditions of the data set, "edges" representing ongoing decision process and "leaf nodes" representing the end of the decision.

The two types of decision trees are: "Classification tree" that is used to classify Data based on the existing data available in the system; "Regression tree" which is used to make predictions for future events based on the existing data in the system. Both of these trees are heavily used in machine learning algorithms. A widely used terminology for decision trees is "Classification and Regression trees" or "CART".

Let's look at how you can build a simple decision tree based on a real-life example.

Step 1: Identify what decision needs to be made, which will serve as a "root node" for the decision tree. For this example, a decision needs to be made on "What would you like to do over the weekend?". Unlike real trees, the decision tree has its roots on top instead of the bottom.

Step 2: Identify conditions or influencing factors for your decision which will serve as "branch nodes" for the decision tree. For this example, conditions could include "would you like to spend the weekend alone or with your friends?" and "how is the weather going to be?".

Step 3: As you answer the conditional questions, you may run into additional conditions that you might have ignored. You will now continue to your final decision by processing all the conditional questions individually, these bifurcations will serve as "edges" of your decision tree.

Step 4: Once you have processed all of the permutations and combinations and eventually made your final decision, that final decision will serve as the "leaf node" of your decision tree. Unlike "branch nodes", there are no further bifurcations possible from a "leaf node".

Here is the graphical representation of your decision for the example above.

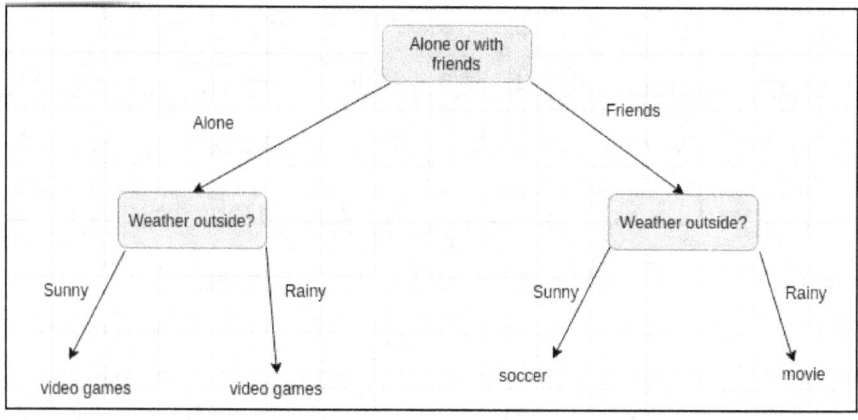

As you would expect from a decision tree, you have obtained a "model representing a set of sequential and hierarchical decisions that ultimately lead to some final decision". This example is at a very high-level to help you develop an understanding of the concept of decision trees. The data science and machine learning decision trees are much more complicated and bigger with hundreds and thousands of branch nodes and edges.

The best tool on the market to visualize and understand decision trees is "Scikit-Learn". Machine learning decisions tree models can be developed using two steps: "Induction" and "Pruning".

Induction

In this step, the decision trees are developed by selecting and modeling all of the sequential and hierarchical decision boundaries based on existing data set. For your ease of understanding, here are 4 high-level steps required to develop the tree:

1. Gather, classify and label the training data set with "feature variables" and "classification or regression output".

2. Identify the best and most cost-effective feature within the training data set that will be used as the point for bifurcating the data.

3. Based on the possible values of the selected "best feature", create subsets of data by bifurcating the data set. These bifurcations will define the "branch nodes" of the decision tree, wherein each node serves as a point of bifurcation based on specific features from the data set.

4. Iteratively develop new tree nodes with the use of data subsets gathered from step 3. These bifurcations will continue until an optimal point is reached, where maximum accuracy is achieved while minimizing the number of bifurcations or nodes.

Pruning

The inherent purpose of decision trees is to support training and self-learning of the system, which often requires overloading of all possible conditions and influencing factors that might affect the final result. To overcome the challenge of setting the correct output for least number of instances per node, developers make a "safe bet" by settling for that "least number" as rather small. This results in a high number of bifurcations on necessary, making for a very complex and large decision tree. This is where "tree pruning" comes into the picture. The verb "prune" literally means "to reduce especially by eliminating superfluous matter". This is the same kind of concept taken from real-life tree pruning and applied to the data science and machine learning decision tree pruning process.

The process of pruning effectively reduces the overall complexity of the decision tree by "transforming and compressing strict and rigid decision boundaries into generalized and smooth boundaries". The number of bifurcations of the decision trees determines the overall complexity of the tree. The easiest and widely used pruning method is reviewing individual branch nodes and evaluating the effect of its removal on the cost function of the decision tree. If the cost function has little to no effect of the removal, then the branch node under review can be easily removed or "pruned".

10. Apriori machine learning algorithm

"Apriori algorithm" is another unsupervised ML algorithm that can produce rules of the association from a specified set of data. "Association rule" simply means if an item X exists then the item Y has a predefined probability of existence. Most rules of association are produced in the format of "IF-THEN" statements. For instance, "IF" someone purchases an iPhone, "THEN" they have most likely purchased an iPhone case as well. The Apriori algorithm can draw these findings by initially observing the number of individuals who purchased an iPhone case while making an iPhone purchase and generating a ratio obtained by dividing the number individuals who bought a new iPhone

(1000) with individuals who also bought an iPhone case (800) with their new iPhones.

The fundamental principles of Apriori ML Algorithm are:

- If a set of events have high frequency of occurrence, then all subsets of that event set will also have high frequency of occurrence.

- If a set of events occur occasionally, then all supersets of the event set of will occur occasionally as well.

Apriori algorithm has wide applicability in the following areas.

"Detecting Adverse Drug Reactions"

"Apriori algorithm" is used to analyze healthcare data such as the drugs administered to the patient, characteristics of each patient, harmful side effects experienced by the patient, the original diagnosis, among others. This analysis generates rules of association that provide insight into the character of the patient and the administered drug that potentially contributed to adverse side effects of the drug.

"Market Basket Analysis"

Some of the leading online e-commerce businesses including "Amazon", use Apriori algorithm to gather insights on products that have high likelihood of being bought together and products that can have an upsell with product promotions and discount offers. For instance, Apriori could be used by a retailer to generate prediction such as customers purchasing sugar and flour have high likelihood of purchasing eggs to bake cookies and cakes.

"Auto-Complete Applications"

The highly cherished auto-complete feature on "Google" is another common Apriori application. When the user starts typing in their keywords for a search, the search engine searches its database, for other related phrases that are usually typed in after a particular word.

11. Support vector machine learning algorithm

"Support Vector Machine" or (SVM) is a type of "supervised ML algorithm", used for "classification" or "regression", where the data set trains SVM on "classes" to be able to classify new inputs. This algorithm operates by classifying the data into various "classes" by discovering a line (hyper-plane) that divides the collection of training data into "classes". Due to availability of various linear hyper-planes, this algorithm attempts to maximize the distance between the different "classes" involved, which is called as "margin maximization". By identifying the line that maximizes the class distance, the likelihood of generalizing apparent to unseen data can be improved.

SVM's can be categorized into two as follows:

- "Linear SVM's" – The training data or classifiers can be divided by a hyper-plane.

- "Non-Linear SVM's" – Unlike linear SVMs, in "non-linear SVM's" the possibility to separate the training data with a hyper-plane is non-existent. For example, the Face Detection training data consists of a group of facial images and another group of non-facial images. The training data is so complicated under such circumstances, that it is difficult to obtain a feature representation of every single vector. It is extremely complex to separate the facial data set linearly from the non-facial data set.

SVM is widely used by different economic organizations for stock market forecasting. For example, SVM is leveraged to compare relative stock performances of various stocks in the same industrial sector. The classifications generated by SVM, aids in the investment-related decision-making process.

The Kernel Trick

The data collected in the real world is randomly distributed and making it too difficult to separate different classes linearly. However, if we can potentially figure out a way to map the data from 2-dimensional space to 3-dimensional space, as shown in the picture below, we will be able to discover a decision surface that separates distinct classes.

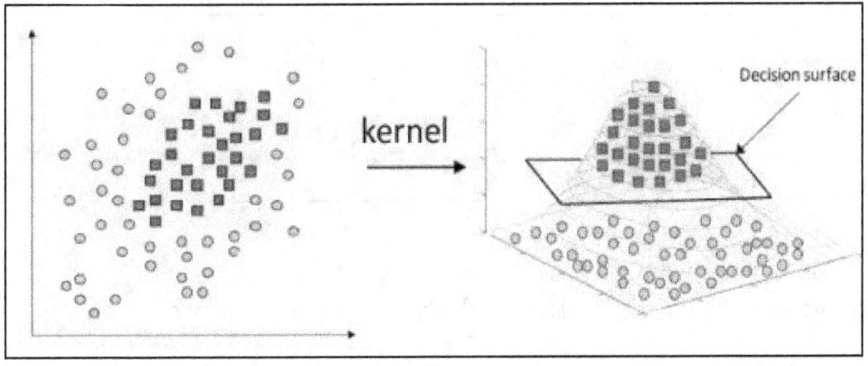

One approach to transforming data like this is mapping all data points to a higher dimension (in this case, 3 dimensions), finding the limit, and making the classification. That works for a limited number of dimensions but computations within given space become increasingly costly when there are a lot of dimensions to deal with. And so the kernel trick comes to the rescue!

The "kernel trick" enables us to function in the original feature space without needing to calculate the data coordinates in a higher-dimensional space. For example, the equation in the picture below has a couple of 3-D data points as 'x' and 'y'.

$$\mathbf{x} = (x_1, x_2, x_3)^T$$
$$\mathbf{y} = (y_1, y_2, y_3)^T$$

Suppose we want to map 'x' and 'y' to 9-dimensional space. To get the outcome, which would be just scalar, we have to do the calculations shown in the picture below. In this case, the computational complexity will be O(n2).

$$\phi(\mathbf{x}) = (x_1^2, x_1x_2, x_1x_3, x_2x_1, x_2^2, x_2x_3, x_3x_1, x_3x_2, x_3^2)^T$$
$$\phi(\mathbf{y}) = (y_1^2, y_1y_2, y_1y_3, y_2y_1, y_2^2, y_2y_3, y_3y_1, y_3y_2, y_3^2)^T$$

$$\phi(\mathbf{x})^T \phi(\mathbf{y}) = \sum_{i,j=1}^{3} x_i x_j y_i y_j$$

However, by using the "kernel function", which is denoted as 'k(x, y)', instead of doing the complex calculations in the 9-dimensional space, the same outcome can be achieved in the 3-dimensional space by calculating the "dot product" of 'x-transpose' and 'y'. In this case, the computational complexity will be O(n).

$$k(\mathbf{x}, \mathbf{y}) = (\mathbf{x}^T \mathbf{y})^2$$
$$= (x_1y_1 + x_2y_2 + x_3y_3)^2$$
$$= \sum_{i,j=1}^{3} x_i x_j y_i y_j$$

In principle, the kernel trick is used to make the transformation of data into higher dimensions much more effective and less costly. The use of the kernel trick is not restricted to the SVM algorithm. The kernel trick can be used with any computations involving the "dot products (x, y)".

Chapter 3: Basics of Data Science Technologies

In the world of technology, Data has described like an "information that a computer is capable of processing and storing". Thanks to our digital lived data has flooded our realities. Our world is drowning with increasing data collected every second, from clicking on a website to monitoring our smartphones and recording our location every second of the day. We could extract alternatives or solutions to our real-world issues that we have not even experienced yet, from the depth of this humongous volume of data. This very process of collecting ideas using mathematical equations and statistics from a measurable set of information can be described as "Data Science". Data scientist's role tends to be very flexible and is often mistaken for the role of a computer scientist or a statistician. Essentially anyone prepared to dig deep into big amounts of data to collect information can be referred to us data science practitioner. For instance, businesses such as "Walmart" keep track and record of customer-made in-store and online purchases providing customized product and service recommendations. Social media platforms such as "Facebook," which enable users to list their "current location", can identify trends of worldwide migration by evaluating the wealth of information that users themselves provide to the platform.

Back in 1960, the earliest recorded use of the word data science was attributed to "Peter Naur", who supposedly used the word "data science" as a replacement for the term computer science and ultimately introduced the word "datalogy". Naur released a book entitled "Concise Survey of Computer Methods" in 1974, with liberal use throughout the book of the term "data science". In 1992, at "The Second Japanese-French Statistics Symposium", the contemporary definition of data science was proposed, with the recognition of the emergence of a new discipline focused primarily on data types, dimensions, and structures.

"Data science continues to evolve as one of the most promising and in-demand career paths for skilled professionals. Today, successful data professionals understand that they must advance past the traditional skills of analyzing large amounts of data, data mining, and programming skills. In order to uncover useful intelligence for their organizations, data scientists must master the full spectrum of the data science life cycle and possess a level of flexibility and understanding to maximize returns at each phase of the process."
– University of California, Berkley

The growing interest of business executives has contributed considerably to the latest increase in the popularity of term data science. A big proportion of journalists and scholarly specialists, however, do not recognize data science as a distinct field of research from the field of statistics. Data science is construed as the popular term for "data mining" and "big data" within the same community. The very definition of the term data science appears to be up for discussion within the tech community. Therefore, largely the field of research a highly versatile skill set including computer programming skills, domain knowledge, proficient statistics abilities and expertise in mathematical algorithms to obtain useful knowledge from large volumes of raw data can be referred to as "data science".

IMPORTANCE AND APPLICATIONS OF DATA SCIENCE

Now that you have a basic understanding of data science, let's look at the significance and applications of data science in our day-to-day lives:

- "Big data and Big Data Analytics" is another 'branch' of data science that organizations use to tackle complicated technological and resource management issues. Later in this book, you will learn more about the concept of big data and big data analytics.

- Data trends have altered dramatically over the last 20 years, indicating a steady rise in unstructured data. It is estimated that "more than 80% of the data we collect will be unstructured by the year 2020".

Conventionally, the data we obtained was primarily structured and could easily be analyzed using the simple business intelligence tools, but as shown in the picture below, unstructured and semi-structured data is on the rise. This, in turn, justifies the creation and use of more powerful and sophisticated analytical tools and technologies, than the current business intelligence tools that are unable to process such humungous volume and variety of data. We need more advanced analytical tools and algorithms that can process and analyze unstructured and semi-structured data to provide useful and actionable insights.

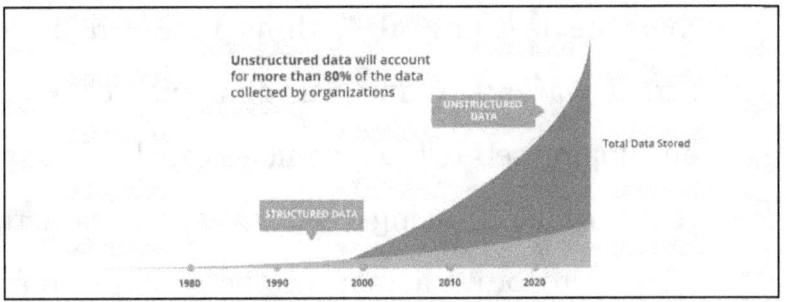

- Companies are always on the verge of understanding their customers' needs better and better. This can now be accomplished by collecting data from current sources such as customer order history, items viewed recently, gender, age and demographics, and using sophisticated analytical tools and algorithms to obtain valuable insights from this data. The use of machine learning algorithms can produce product suggestions for individual clients with greater precision. The "smart" customer is always searching for the most engaging and enhanced user experience, so these analytical tools and algorithms can be used by businesses to gain a competitive advantage and grow their company.

- Data science has made it possible to use sophisticated machine learning algorithms that can be applied across various industrial domains. For instance, developing self-driving vehicles capable of gathering real-time data, using their advanced cameras and sensors to build a map of their environment and make choices about car velocity and other driving maneuvers.

- Data science is used extensively in "predictive analysis". Weather forecasting, for instance, needs data collection and analysis from a multitude of sources including satellites, radars, and aircraft to construct data models that can effectively predict the occurrence of natural disasters such as hurricanes, tornadoes and flash floods.

- The ability of Data science to evaluate the problems facing companies across the industrial spectrum, such as healthcare, travel, finance, retail, and e-commerce, has considerably led to its growing popularity among business leaders.

BIG DATA AND BIG DATA ANALYTICS

In 2001, Gartner defined Big Data as "Data that contains greater variety arriving in increasing volumes and with ever-higher velocity"; which led to the formulation of initial "Three Vs of Big Data". Big data relates to an endlessly flooding avalanche of structured and unstructured data from a multitude of infinite sources of information. These information sets are too big to be analyzed using traditional analytical tools and technologies, but they have a wealth of precious insights hidden in the depth.

The origin of "Big Data" has been traced back to the 1960s and 1970s, when the "Third Industrial Revolution" had just begun to kick in and the growth of relational database had started along with the construction of data centers. But with the high ease of accessibility of free search engines such as "Google" and "Yahoo", free internet entertainment services such as "YouTube" and social media platforms such as "Facebook" and "Twitter", the concept of big data has lately taken center stage. Companies started acknowledging the incredible amount of user data generated through these platforms and services in 2005, and in the same year, the open-source framework called "Hadoop" was developed to collect and analyze these large data dumps available to these companies. A non-relational or distributed database called "NoSQL" began to gain popularity during the same era, because of its capacity to store and extract unstructured data. "Hadoop" enabled businesses to operate with big data at a comparatively low cost and with great ease.

Today, with the rise of cutting-edge technology, not only humans, but we have developed data-generating machines. Smart device technologies such as "Internet of Things" (IoT) and "System Internet" (IoS) have skyrocketed the volume of big data. Our everyday household objects and smart devices are linked to the Internet and are capable of tracking and recording our usage patterns as well as our interactions with these products and directly feeding all this information into big data. The emergence of machine learning technology has further enhanced the daily amount of information produced. It is estimated that "1.7 MB of information per second per person will be produced by 2020". As big data continues to expand, there are still many horizons to be crossed to achieve maximum usability.

"The importance of big data doesn't revolve around how much data you have, but what you do with it. You can take data from any source and analyze it to find answers that enable 1) cost reductions, 2) time reductions, 3) new product development and optimized offerings, and 4) smart decision making."

- SAS

"THE VS OF BIG DATA"

Volume – The "volume" of the given data set must be substantially larger than traditional data sets to be classified as big data. These data sets consist mainly of unstructured data with limited structured and semi-structured data. Input sources such as web pages, search history, mobile apps, and social media platforms can provide data with unknown significance or unstructured data. The company's size and client base is generally proportional to the volume of data the firm can easily obtain.

Velocity – The speed at which data can be collected and the processed refers to the velocity of big data. The combination of on-premise and cloud-based servers is increasingly being used by companies to improve the pace and storage of their data collection. Modern-day "Smart Products and Devices" utilize real-time access to customer information, to provide them with more engaging and improved user experience.

Variety – Conventionally, a data set would comprise of mostly structured data with a small quantity of unstructured and semi-structured data, but the emergence of big data has given rise to new unstructured data types such as video, text, audio that require advanced tools and technologies to clean and process for extraction of meaningful and actionable insights.

Veracity – Veracity is another "V" to be regarded for big data analytics. This relates to the "reliability or quality" of the data. For instance, social media platforms such as "Facebook" and "Twitter" with blogs and posts, flooded with hashtags and acronyms and all types of typing errors can considerably decrease the reliability and accuracy of the data sets.

Value – Data has developed as an inherent currency of its own. Like traditional currencies, big data's ultimate value is directly proportional to the usefulness and quality of insights collected from it.

IMPORTANCE AND APPLICATIONS OF BIG DATA/BIG DATA ANALYTICS

To gain accurate and reliable information from a data set, it is very essential to have a full data set and that can be accomplished with the use of big data technology. The more data we have, the more details and insights can be obtained from it. The future of big data is very promising for obtaining a 360 view of a problem and its underlying solutions. Here are some examples of the application of big data.

Product development – Big data is progressively being used by large and small e-commerce companies to comprehend client requirements and expectations. Companies can create predictive models for launching new products and services using the main features of their past and current products and services and producing a model describing the connection between those features and the business achievement of those products and services. For instance, "Procter & Gamble," a major fast-moving commercial goods company, widely utilizes large information collected from social media websites, test markets, and focus groups to prepare for their new product launch.

Predictive maintenance – A large volume of unstructured data such as error messages, log entries, and ordinary machine temperature must be evaluated along with accessible structured data, such as machine make, model and production year, to avoid potential mechanical and equipment failures. By evaluating this big data set using the necessary analytical instruments, businesses can extend their equipment's shelf life by preparing ahead of time for scheduled maintenance and predicting potential mechanical failures in the future.

Customer experience – The "smart customer" is conscious of all the advances in technology and is only faithful to the most engaging and enhanced user experience available. This has led to a race of the companies to provide distinctive customer experiences analyzing the information collected from customer interactions with the goods and services of the company. Providing personalized recommendations and discount offers, companies can lower the churn rate of their customers and efficiently convert potential leads into paying buyers.

Fraud and compliance – Big data enables identification of data patterns and supports deterrence of possibly fraudulent transactions in the future, by defining data patterns and evaluating historical trends from past fraudulent transactions. Banks, financial institutions and internet payment services such as "PayPal" are continually tracking and collecting information on customer transactions in an attempt to avoid fraud.

Operational effectiveness – Using predictive analysis of big data, companies can learn and anticipate future demand and product trends by evaluating manufacturing capacity, customer feedback, and information on top-selling items and product returns, to enhance decision-making and generate goods that are in line with current market trends.

Machine learning – For a machine to be able to learn and train on its own it needs a huge quantity of data, i.e. Big Data. A solid training set with structured, semi-structured and unstructured data will help the machine develop a multi-dimensional view of the real world and the problem to be solved.

Drive innovation – By studying and understanding the interactions between individuals and their electronic devices as well as the manufacturers of these devices, companies can create enhanced and innovative goods by examining current product trends and meeting client expectations.

DATA MINING

Data mining technology is defined as "the process of exploring and analyzing large volumes of data to gather meaningful patterns and rules". Data mining comes under the umbrella of data science and is used to construct, artificial intelligence-based machine learning models, for instance, search engine algorithms. Although the procedure of "digging through data" to uncover hidden patterns and predict future events has been around for a while and known as "database knowledge discovery," the term "data mining" was coined as recently as the 1990s.

Data mining consists of three foundational and highly intertwined science disciplines, namely "statistics" (the mathematical study of data relationships), "machine learning algorithms" (algorithms that can be trained with an inherent capability to learn) and "artificial intelligence" (machines that can display human-like intelligence). Data Mining technology has evolved with the introduction of Big data analytics to keep up with the "unlimited potential of big data" and affordable computing power. Using advanced processing speed and power of contemporary computing devices, the once considered tedious, labor-intensive and time-consuming operations have been successfully automated.

"Data mining is the process of finding anomalies, patterns and correlations within large data sets to predict outcomes. Using a broad range of techniques, you can use this information to increase revenues, cut costs, improve customer relationships, reduce risks and more."

– SAS

According to SAS, "unstructured data alone makes up 90% of the digital universe". This big data avalanche does not necessarily ensure more knowledge and understanding. Using data mining technology enables all redundant and unnecessary data noise to be filtered out, so to garner the understanding of applicable information used in the immediate decision-making process.

Importance and applications of Data mining

The applications of data mining techniques are widespread, ranging from retail pricing and promotions to loan risk evaluation by financial organizations and banks. Companies across the industrial spectrum can benefit from the applications of data mining technology. Here are some examples of industrial applications of data mining technology:

Healthcare bio-informatics

Statistical models are used by healthcare practitioners to estimate the probability of patients suffering from one or more health condition considering the risk factors. Genetically transmitted illnesses can be avoided or mediated by modeling the genetic, family and demographic data of the patient from the onset of declining health condition. There is a shortage of healthcare experts in developing countries, so assisted diagnoses and patient prioritization are very critical. Models based on data mining have lately been implemented in these nations to assist with prioritization of patients before healthcare practitioners can achieve and administer therapy in these nations.

Credit Risk Management

Financial organizations and banks implement data mining based tools and models that predict the probability of a prospective credit card customer failing to make their credit payments on time and determining the suitable loan limit that the customer may be eligible for. These data mining models collect and extract information from a multitude of input sources including personal information, customer's financial history, and demographics, among others. The model then offers the interest rate to be obtained from the customer by the organization or bank depending on the assessed risk of lending. For instance, the applicant's credit score is taken into account by data mining models and elevated interest rates are provided to people with low credit score.

Spam filtering

A lot of email clients like "Google mail" and "Yahoo mail" depend on data mining tools to detect and flag email spam and malware. The data mining tools generate insights and understanding that can be used to develop improved safety measures and tools by analyzing hundreds and thousands of shared features of spam and malware. Not only are these applications able to detect spam, but they are also very effective in categorizing spam messages and storing them in a distinct directory, so they never enter the user's inbox.

Marketing

Retail businesses need to know their customer demands and expectations to the dot. Businesses can evaluate customer-related information such as order history, demographics, gender and age by using data mining tools to collect useful customer insights and segment them into organizations based on shared shopping characteristics. Then companies develop distinctive advertising policies and campaigns to target particular groups such as discount offers and product promotions.

Sentiment analysis

Companies can analyze their data from all their social media platforms using a method called "text mining" to know their customer base's "sentiment". This process of understanding emotions of a large group of people towards a specific subject is called "sentiment analysis" and can be performed using data mining tools. Using pattern recognition technology, input data from social media platforms and other associated public content websites are gathered using "text mining" technology and data trends identify that feed into the general knowledge of the subject matter. The "natural language processing" method can be used to further dive into this data to comprehend the human language in a particular context.

Qualitative data mining

The "text mining" method can also be used to conduct qualitative research and obtain insights from large quantities of unstructured data. A recent research study by "University of California, Berkeley" disclosed the use of data mining models in child welfare program studies.

Product recommendation systems

Advance "recommendation systems" are like online retailers' bread and butter. To achieve a competitive edge in the industry, the use of predictive customer behavior analysis is increasing among small and large online businesses. Some of the biggest e-commerce companies, including "Amazon", "Macy's" and "Nordstrom", have invested millions of bucks in developing their proprietary data mining models to forecast market trends and offer their customers more engaging and enhanced user experience. The on-demand entertainment giant "Netflix" purchased an algorithm worth more than a million dollars to improve the precision of their video recommendation system, which allegedly improved the accuracy of the recommendation for "Netflix" by more than 8%.

Artificial Intelligence

Psychology professionals classified "human intelligence" as a powerful combination of a multitude of mental skills including the capacity to learn from life experiences, capacity to adapt to evolving settings, abstract ideas, logical reasoning, perception, language capacity, and problem-solving skills. Not only have human beings experimented with livestock to detect indications of intelligence for the greater purpose, but we have also experimented with devices to impart them as human intelligence. And that's where the "Artificial Intelligence" concept came into being.

The pioneering British computer scientist, Alan Turing, laid the foundation for artificial intelligence technology in the mid-20th century, with the development of an abstract machine with a scanner and limitless memory capable of altering and enhancing his programming that implied a learning capacity. Today, this machine serves as the basis for the development and improvement of the modern computer and is known as "Turing machine." Eventually, in 1956, the term "artificial intelligence" or "AI" was coined and can be defined as the science of the development of machines and computers controlled and operated by human beings, but capable of mirroring and manifesting human intelligence to achieve this.

The ability to learn, reason and perceive are considered as the three fundamental goals of Artificial Intelligence. Some of the core human mental abilities that researchers are aspiring to mimic in computers and machines are:

1. **Knowledge** – The machines require a big amount of information to comprehend and process the world as it is. For the development of artificial knowledge engineering, seamless access to information objects, categories, characteristics and interactions that are stored and managed using data storage is critical.

2. **Learning** – The Evergreen technique of testing and error tends to be the easiest type of teaching for artificial intelligence technology.

3. **Problem Solving** – "The systematic process to reach a predefined goal or solution by searching through a range of possible actions" can be defined as problem-solving.

4. **Reasoning** – The capacity to draw inferences is referred to as the act of reasoning in compliance with the scenario at hand. Two commonly recognized types of reasoning are "deductive reasoning" (assumes the conclusion is true if the assumption is true) and "inductive logic" (even if the assumption is true, the conclusion might or might not be true). One of the most challenging challenges in the growth and promotion of artificial intelligence technology is the implementation of "true reasoning".

5. **Perception** – Perception relates to the process of creating a 3-D view of an object using different sensory organs and is directly influenced by the surrounding environment. Artificial perception has already created self-driving cars and the ability to collect and supply products.

Chapter 4: Machine Learning Library "Scikit-Learn" 101

Machine learning libraries are sensitive routines and functions that are written in any given language. Software developers require a robust set of libraries to perform complex tasks without needing to rewrite multiple lines of code. Machine learning is largely based on mathematical optimization, probability, and statistics.

Python is the language of choice in the field of machine learning credited to consistent development time and flexibility. It is well suited to develop sophisticated models and production engines that can be directly plugged into production systems. One of its greatest assets being an extensive set of libraries that can help researchers who are less equipped with developer knowledge to easily execute machine learning.

"Scikit-Learn" has evolved as the gold standard for machine learning using Python, offering a wide variety of "supervised" and "unsupervised" ML algorithms. It is touted as one of the most user-friendly and cleanest machine learning libraries to date. For example, decision trees, clustering, linear and logistics regressions and K-means. Scikit-learn uses a couple of basic Python libraries: NumPy and SciPy and adds a set of algorithms for data mining tasks including classification, regression, and clustering. It is also capable of implementing tasks like feature selection, transforming data and ensemble methods in only a few lines.

In 2007, David Cournapeau developed the foundational code of "Scikit-Learn" as part of a "Summer of Code" project for "Google". Scikit-learn has become one of Python's most famous open-source machine learning libraries since its launch in 2007. But it wasn't until 2010 that Scikit-Learn was released for public use. Scikit-Learn is an open-sourced and BSD licensed, data mining and data analysis tool used to develop supervise and unsupervised machine learning algorithms build on Python. Scikit-learn offers various ML algorithms such as "classification", "regression", "dimensionality reduction", and "clustering". It also offers modules for feature extraction, data processing, and model evaluation.

Designed as an extension to the "SciPy" library, Scikit-Learn is based on "NumPy" and "matplotlib", the most popular Python libraries. NumPy expands Python to support efficient operations on big arrays and multidimensional matrices. Matplotlib offers visualization tools and science computing modules are provided by SciPy. For scholarly studies, Scikit-Learn is popular because it has a well-documented, easy-to-use and flexible API. Developers can utilize Scikit-Learn for their experiments with various algorithms by only altering a few lines of the code. Scikit-Learn also provides a variety of training datasets, enabling developers to focus on algorithms instead of data collection and cleaning. Many of the algorithms of Scikit-Learn are quick and scalable to all but huge datasets. Scikit-learn is known for its reliability and automated tests are available for much of the library. Scikit-learn is extremely popular with beginners in machine learning to start implementing simple algorithms.

PREREQUISITES FOR APPLICATION OF SCIKIT-LEARN LIBRARY

The Scikit-Learn library is based on the SciPy (Scientific Python), which needs to be installed before using SciKit-Learn. This stack involves the following:

NumPy (Base n-dimensional array package)

"NumPy" is the basic package with Python to perform scientific computations. It includes among other things: "a powerful N-dimensional array object; sophisticated (broadcasting) functions; tools for integrating C/C++ and Fortran code; useful linear algebra, Fourier transform, and random number capabilities". NumPy is widely reckoned as an effective multi-dimensional container of generic data in addition to its apparent scientific uses. It is possible to define arbitrary data types. This enables NumPy to integrate with a broad variety of databases seamlessly and quickly. The primary objective of NumPy is the homogeneity of a multidimensional array. It consists of an element table (generally numbers), all of which are of the same sort and are indicated by tuples of non-negative integers. The dimensions of NumPy are called "axes" and array class is called "ndarray".

Matplotlib (Comprehensive 2D/3D plotting)

"Matplotlib" is a 2-dimensional graphic generation library from Python that produces high-quality numbers across a range of hardcopy formats and interactive environments. The "Python script", the "Python", "IPython shells", the "Jupyter notebook", the web app servers, and select user interface toolkits can be used with matplotlib. Matplotlib attempts to further simplify easy tasks and make difficult tasks feasible. With only a few lines of code, you can produce tracks, histograms, scatter plots, bar graphs, error graphs, etc.

A MATLAB-like interface is provided for easy plotting of the Pyplot Module, especially when coupled with IPython. As a power user, you can regulate the entire line styles, fonts properties, and axis properties, through an object-oriented interface or a collection of features similar to the one provided to MATLAB users.

SciPy (Fundamental library for scientific computing)

SciPy is a "collection of mathematical algorithms and convenience functions built on the NumPy extension of Python", capable of adding more impact to interactive Python sessions, by offering high-level data manipulation and visualization commands and courses for the user. An interactive Python session with SciPy becomes an environment that rivals data processing and system prototyping technologies including "MATLAB, IDL, Octave, R-Lab, and SciLab".

Another advantage of developing "SciPy" on Python, is the accessibility of a strong programming language in the development of advanced programs and specific apps.

Scientific apps using SciPy benefit from developers around the globe developing extra modules in countless software landscape niches. Everything produced has been made accessible to the Python programmer, from database subroutines and classes as well as "parallel programming to the web". These powerful tools are provided along with the "SciPy" mathematical libraries.

IPython (Enhanced interactive console)

"IPython (Interactive Python)" is an interface or command shell for interactive computing using a variety of programming languages. "IPython" was initially created exclusively for Python, which supports introspection, rich media, shell syntax, tab completion, and history. Some of the functionalities provided by IPython include: "interactive shells (terminal and Qt-based); browser-based notebook interface with code, text, math, inline plots and other media support; support for interactive data visualization and use of GUI tool kits; flexible interpreters that can be embedded to load into your own projects; tools for parallel computing".

SymPy (Symbolic mathematics)

Developed by Ondřej Čertík and Aaron Meurer, SymPy is "an open-source Python library for symbolic computation". It offers algebra computing abilities to other apps, as a stand-alone app and/or as a library as well as live on the internet applications with "SymPy Live" or "SymPy Gamma". "SymPy" is easy to install and test, because it is completely developed in Python boasting limited dependencies. SymPy involves characteristics ranging from calculus, algebra, discrete mathematics, and quantum physics to fundamental symbolic arithmetic. The outcome of the computations can be formatted as "LaTeX" code. In combination with a straightforward, expandable code base in a widespread programming language, the ease of access provided by SymPy makes it a computer algebra system with comparatively low entry barrier.

Pandas (Data structures and analysis)

Pandas provide highly intuitive and user-friendly high-level data structures. Pandas has achieved popularity in the machine learning algorithm developer community, with built-in techniques for data aggregation, grouping, and filtering as well as results of time series analysis. The Pandas library has two primary structures: one-dimensional "Series" and two-dimensional "Data Frames."

Seaborn (data visualization)

Seaborn is derived from the Matplotlib Library and an extremely popular visualization library. It is a high-level library that can generate specific kind of graph including heat map, time series, and violin plots.

INSTALLING SCIKIT-LEARN

The latest version of Scikit-Learn can be found on "Scikit-Learn.org" and requires "Python (version >= 3.5); NumPy (version >= 1.11.0); SciPy (version >= 0.17.0); joblib (version >= 0.11)". The plotting capabilities or functions of Scikit-learn start with "plot_" and require "Matplotlib (version >= 1.5.1)". Certain Scikit-Learn examples may need additional applications: "Scikit-Image (version >= 0.12.3), Pandas (version >= 0.18.0)".

With the previous installation of "NumPy" and "SciPy", the best method of installing Scikit-Learn is using "pip: pip install -U scikit-learn" or "conda: conda install scikit-learn".

One must make sure that "binary wheels" are utilized when using pip and that "NumPy" and "SciPy" have not been recompiled from source, which may occur with the use of specific OS and hardware settings (for example, "Linux on a Raspberry Pi"). Developing "NumPy" and "SciPy" from source tends to be complicated (particularly on Windows), therefore, they need to be setup carefully making sure optimized execution of linear algebra routines is achievable.

APPLICATION OF MACHINE LEARNING USING SCIKIT-LEARN LIBRARY

To understand how Scikit-Learn library is used in development of machine learning algorithm, let us use the "Sales_Win_Loss data set from IBM's Watson repository" containing data obtained from sales campaign of a wholesale supplier of automotive parts. We will build a machine learning model to predict which sales campaign will be a winner and which will incur a loss.

The data set can be imported using Pandas and explored using Pandas techniques such as "head(), tail() and dtypes()". The plotting techniques from "Seaborn" will be used to visualize the data. To process the data Scikit-Learn's "preprocessing.LabelEncoder()" will be used and "train_test_split()" to divide the data set into a training subset and testing subset.

To generate predictions from our data set, three different algorithms will be used namely, "Linear Support Vector Classification and K-nearest neighbors classifier". To compare the performances of these algorithms Scikit-Learn library technique "accuracy_score" will be used. The performance score of the models can be visualized using Scikit-Learn and "Yellowbrick" visualization.

IMPORTING THE DATA SET

To import the "Sales_Win_Loss data set from IBM's Watson repository", the first step is importing the "Pandas" module using "*import pandas as pd*".

Then we leverage a variable url as " *https://community.watsonanalytics.com/wp content/uploads/2015/04/WA_Fn-UseC_-Sales-Win-Loss.csv* " to store the URL from which the data set will be downloaded.

Now, *"read_csv() as sales_data = pd.read_csv(url)"* technique will be used to read the above "csv or comma-separated values" file, which is supplied by the Pandas module. The csv file will then be converted into a Pandas data framework, with the return variable as "*sales_data*", where the framework will be stored.

For new 'Pandas' users, the *"pd.read csv()"* technique in the code mentioned above will generate a tabular data structure called "data framework", where an index for each row is contained in the first column, and the label/name for each column in the first row are the initial column names acquired from the data set. In the above code snippet, the *"sales data"* variable results in a table depicted in the picture below.

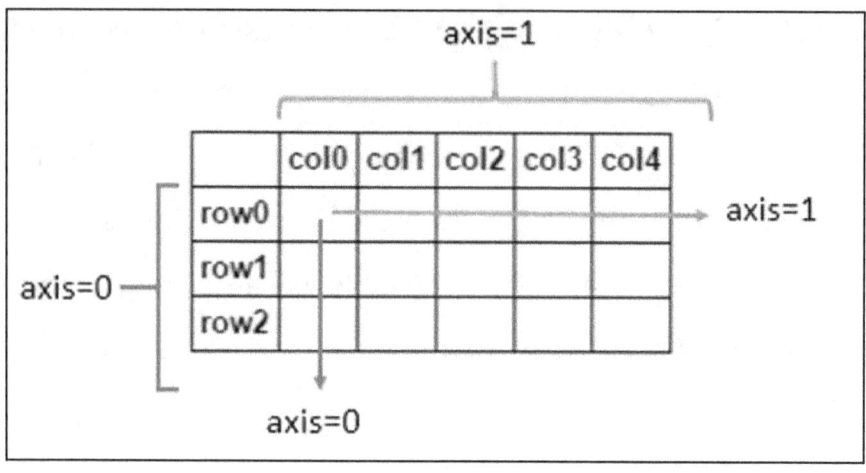

In the diagram above, the "row0, row1, row2" represent individual record index and the "col0, col1, col2" represent the names for individual columns or features of the data set.

With this step, you have successfully stored a copy of the data set and transformed it into a "Pandas" framework!

Now, using the *"head() as Sales_data.head()"* technique the records from the data framework can be displayed as shown below to get a "feel" of the information contained in the data set.

	opportunity number	supplies subgroup	supplies group	region	route to market	elapsed days in sales stage	opportunity result
0	1641984	Exterior Accessories	Car Accessories	Northwest	Fields Sales	76	Won
1	1658010	Exterior Accessories	Car Accessories	Pacific	Reseller	63	Loss
2	1674737	Motorcycle Parts	Performance & Non-auto	Pacific	Reseller	24	Won
3	1675224	Shelters & RV	Performance & Non-auto	Midwest	Reseller	16	Loss

DATA EXPLORATION

Now that we have our copy of the data set which has been transformed it into a "Pandas" data frame, we can quickly explore the data to understand what information can tell can be gathered from it and accordingly to plan a course of action.

In any ML project, data exploration tends to be a very critical phase. Even a fast data set exploration can offer us significant information that could be easily missed otherwise, and this information can propose significant questions that we can then attempt to answer using our project.

Some third-party Python libraries will be used here to assist us with the processing of the data so that we can efficiently use this data with the powerful algorithms of Scikit-Learn. The same *"head()"* technique that we used to see some initial records of the imported data set in the earlier section can be used here. As a matter of fact, *"(head)"* is effectively capable of doing much more than displaying data record and customize the "head()" technique to display only selected records with commands like *"sales_data.head(n=2)"*. This command will selectively display the first 2 records of the data set. At first sight, it's obvious, that columns such as "Supplies Group" and "Region" contain string data, while columns such as "Opportunity Result", "Opportunity Number" etc. are comprised of integer values. It can also be seen that there are unique identifiers for each record in the' Opportunity Number' column.

Similarly, to display select records from the bottom of the table, the *"tail() as sales_data.tail()"* can be used.

To view the different data types available in the data set, the Pandas technique *"dtypes() as sales_data.dtypes"* can be used. With this information, the data columns available in the data framework can be listed with their respective data types. We can figure out, for example, that the column "Supplies Subgroup" is an "object" data type and that the column "Client Size By Revenue" is an "integer data type". So, we have an understanding of columns that either contain integer values or string data.

DATA VISUALIZATION

At this point we are through with basic data exploration steps, so we will not attempt to build some appealing plots to portray the information visually and discover other concealed narratives from our data set.

Of all the available Python libraries providing data visualization features; "Seaborn" is one of the best available options so we will be using the same. Make sure that python plots module provided by "Seaborn" has been installed on your system and ready to be used. Now follow the steps below generate the desired plot for the data set:

Step 1 - Import the "Seaborn" module with the command *"import seaborn as sns"*.

Step 2 - Import the "Matplotlib" module with command *"import matplotlib.pyplot as plt"*.

Step 3 - To set the "background colour" of the plot as white, use command *"sns.set(style="whitegrid", color_codes=True)"*.

Step 4 - To set the "plot size" for all plots, use command *"sns.set(rc={'figure.figsize':(11.7,8.27)})"*.

Step 5 – To generate a "countplot", use command "*sns.countplot('Route To Market',data=sales_data,hue = 'Opportunity Result')*".

Step 6 – To remove the top and bottom margins, use command "*sns.despine(offset=10, trim=True)*".

Step 7 – To display the plot, , use command "*plotplt.show()*".

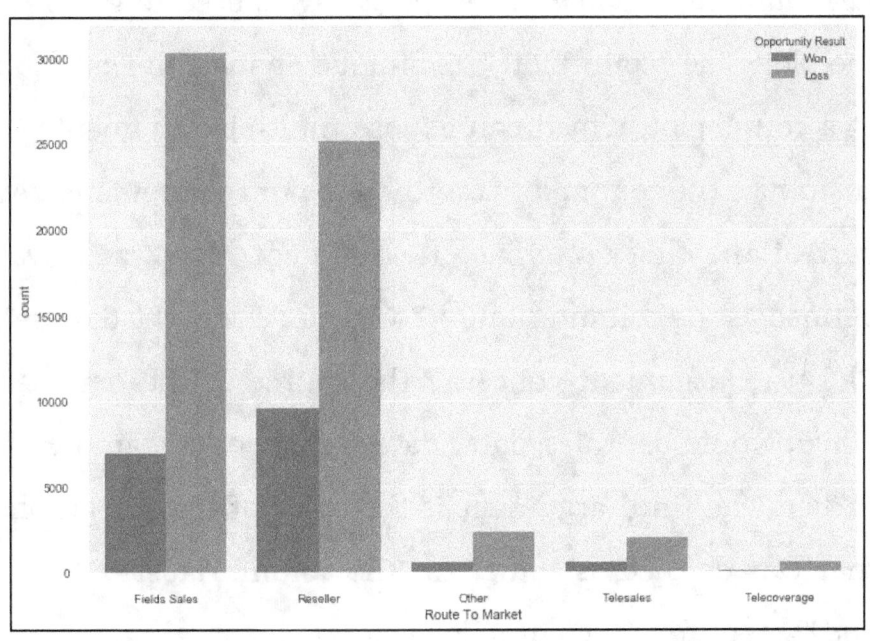

Quick recap - The "Seaborn" and "Matplotlib" modules were imported first. Then the *"set()"* technique was used to define the distinct characteristics for our plot, such as plot style and color. The background of the plot was defined to be white using the code snippet *"sns.set(style= "whitegrid", color codes= True)"*. Then the plot size was define using command *"sns.set(rc={'figure.figsize':(11.7,8.27)})"* that define the size of the plot as "11.7px and 8.27px".

Next the command *"sns.countplot('Route To Market',data= sales data, hue='Opportunity Result')"* was used to generate the plot. The "countplot()" technique enables the creation of a count plot, which can expose multiple arguments to customize the count plot according to our requirements. As part of the first *"countplot()"* argument, the X-axis was defined as the column "Route To Market" from the data set. The next argument concerns the source of the data set, which would be "sales_data" data framework we imported earlier. The third argument is the color of the bar graphs that was defined as "blue" for the column labeled "won" and "green" for the column labeled "loss".

DATA PRE-PROCESSING

By now you should have a clear understanding of what information is available in the data set. From the data exploration step, we established that majority of the columns in our data set are "string data", but "Scikit-Learn" can only process numerical data. Fortunately, the Scikit-Learn library offers us many ways to convert string data into numerical data, for example, *"LabelEncoder()"* technique. To transform categorical labels from the data set such as "won" and "loss" into numerical values, we will use the *"LabelEncoder()"* technique.

Let's look at the pictures below to see what we are attempting to accomplish with the *"LabelEncoder()"* technique. The first image contains one column labeled "color" with three records namely, "Red", "Green" and "Blue". Using the *"LabelEncoder()"* technique, the record in the same "color" column can be converted to numerical values, as shown in the second image.

	Color
0	Red
1	Green
2	Blue

	Color
0	1
1	2
2	3

Let's begin the real process of conversion now. Using the *"fit transform()"* technique given by *"LabelEncoder()"*, the labels in the categorical column like "Route To Market" can be encoded and converted to numerical labels comparable to those shown in the diagrams above. The function *"fit transform()"* requires input labels identified by the user and consequently returns encoded labels.

To know how the encoding is accomplished, let's go through an example quickly. The code instance below constitutes string data in form of a list of cities such as ["paris", "paris", "tokyo", "amsterdam"] that will be encoded into something comparable to "[2, 2, 1,3]".

Step 1 - To import the required module, use the command *"from sklearn import preprocessing"*.

Step 2 – To create the Label encoder object, use command *"le = preprocessing.LabelEncoder()"*.

Step 3 – To convert the categorical columns into numerical values, use command:

"encoded_value = le.fit_transform(["paris", "paris", "tokyo", "amsterdam"])"

"print(encoded_value) [1 1 2 0]"

And there you have it! We just converted our string data labels into numerical values. The first step was importing the preprocessing module that offers the *"LabelEncoder()"* technique. Followed by development of an object representing the *"LabelEncoder()"* type. Then the *"fit_transform()"* function of the object was used to distinguish between distinct classes of the list ["paris", "paris", "tokyo", "amsterdam"] and output the encoded values of "[1 1 20]".

Did you observe that the *"LabelEncoder()"* technique assigned the numerical values to the classes in alphabetical order according to the initial letter of the classes, for example "(a)msterdam" was assigned code "0", "(p)aris" was assigned code "1" and "(t)okyo" was assigned code "2".

CREATING TRAINING AND TEST SUBSETS

To know the interactions between distinct characteristics and how these characteristics influence the target variable, a ML algorithm must be trained on a collection of information. We need to split the complete data set into two subsets to accomplish this. One subset will serve as the training data set, which will be used to train our algorithm to construct machine learning models. The other subset will serve as the test data set, which will be used to test the accuracy of the predictions generate by the machine learning model.

The first phase in this stage is the separation of feature and target variables using the steps below:

Step 1 – To select data excluding select columns, use command *"select columns other than 'Opportunity Number', 'Opportunity Result'cols = [col for col in sales_data.columns if col not in ['Opportunity Number','Opportunity Result']]"*.

Step 2 – To drop these select columns, use command *"dropping the 'Opportunity Number'and 'Opportunity Result' columns*

data = sales_data[cols]".

Step 3 – To assign the Opportunity Result column as "target", use command *"target = sales_data['Opportunity Result']*

data.head(n=2)".

The "Opportunity Number" column was removed since it just acts as a unique identifier for each record. The "Opportunity Result" contains the predictions we want to generate, so it becomes our "target" variable and can be removed from the data set for this phase. The first line of the above code will select all the columns except "Opportunity Number" and "Opportunity Result" in and assign these columns to a variable "cols". Then using the columns in the "cols" variable a new data framework was developed. This is going to be the "feature set". Next, the column "Opportunity Result" from the *"sales_data"* data frame was used to develop a new data framework called "target".

The second phase in this stage concerns the separation of the date frameworks into training and testing subsets using the steps below. Depending on the data set and desired predictions, it needs to be split into training and testing subset accordingly. For this exercise, we will use 75% of the data as a training subset and the rest 25% will be used for the testing subset. We will leverage the *"train_test_split()"* technique in "Scikit-Learn" to separate the data using steps and code as below:

Step 1 – To import the required module, use the command *"from sklearn.model_selection import train_test_split"*.

Step 2 – To separate the data set, use command *"split data set into train and test setsdata_train, data_test, target_train, target_test = train_test_split(data,target, test_size = 0.30, random_state = 10)"*.

With the code above, the *"train_test_split"* module was first imported, followed by the use of *"train_test_split()"* technique to generate *"training subset (data_train, target_train)"* and *"testing subset (data_test, data_train)"*. The *"train_test_split()"* technique's first argument pertains to the features that were divided in the preceding stage, the next argument relates to the target ("Opportunity Result"). The third "test size" argument is the proportion of the data we wish to divide and use as a testing subset. We are using 30% for this example, although it can be any amount. The fourth 'random state' argument is used to make sure that the results can be reproduced every time.

BUILDING THE MACHINE LEARNING MODEL

The "machine_learning_map" provided by Scikit-Learn is widely used to choose the most appropriate ML algorithm for the data set. For this exercise, we will be using "Linear Support Vector Classification" and "K-nearest neighbors classifier" algorithms.

Linear Support Vector Classification

"Linear Support Vector Classification" or "Linear SVC" is a sub-classification of "Support Vector Machine (SVM)" algorithm, which we have reviewed in chapter 2 of this book titled "Machine Learning Algorithms". Using Linear SVC, the data can be divided into different planes so the algorithm can identify the optimal group structure for all the data classes.

Here are the steps and code for this algorithm to build our first ML model:

Step 1 – To import the required modules, use commands *"from sklearn.svm import LinearSVC"* and *"from sklearn.metrics import accuracy_score"*.

Step 2 – To develop an LinearSVC object type, use command *"svc_model = LinearSVC(random_state=0)"*.

Step 3 – To train the algorithm and generate predictions from the testing data, use command *"pred = svc_model.fit(data_train, target_train).predict(data_test)"*.

Step 4 – To display the model accuracy score, use command *"print ('LinearSVC accuracy:', accuracy_score(target_test, pred, normalize = True))"*.

With the code above, the required modules were imported in the first step. We then developed a type of Linear SVC using *"svc_model"* object with *"random_state"* as '0'. The *"random_state"* command instructs the built-in random number generator to shuffle the data in a particular order. In step 3, the "Linear SVC" algorithm is trained on the training data set and subsequently used to generate predictions for the target from the testing data. The *"accuracy_score()"* technique was used in the end to verify the "accuracy score" of the model, which could be displayed as "LinearSVC accuracy: 0.777811004785", for instance.

K-nearest Neighbors Classifier

The "k-nearest neighbors(k-NN)" algorithm is referred to as "a non-parametric method used for classification and regression in pattern recognition". In cases of classification and regression, "the input consists of the nearest k closest training examples in the feature space". K-NN is a form of "instance-based learning", or "lazy learning", in which the function is only locally estimated and all calculations are delayed until classification. The output is driven by the fact, whether the classification or regression method is used for k-NN:

- "k-nearest neighbors classification" - The "output" is a member of the class. An "object" is classified by its neighbors' plurality vote, assigning the object to the most prevalent class among its nearest "k-neighbors", where "k" denotes a small positive integer. If k= 1, the "object" is simply allocated to the closest neighbor's class.

- "k-nearest neighbors regression" - The output is the object's property value, which is computed as an average of the k-nearest neighbors values.

A helpful method for both classification and regression can be assigning weights to the neighbors' contributions, to allow closer neighbors to make more contributions in the average, compared to the neighbors located far apart. For instance, a known "weighting scheme" is to assign each neighbor a weight of "$1/d$", where "d" denotes the distance from the neighbor. The neighbors are selected from a set of objects for which the "class" (for "k-NN classification") or the feature value of the "object" (for "k-NN regression") is known.

Here are the steps and code for this algorithm to build our next ML model:

Step 1 – To import required modules, use the command *"from sklearn.neighbors import KNeighborsClassifier"* and *"from sklearn.metrics import accuracy_score"*.

Step 2 – To create object of the classifier, use command *"neigh = KNeighborsClassifier(n_neighbors=3)"*.

Step 3 – To train the algorithm, use command *"neigh.fit(data_train, target_train)"*.

Step 4 – To generate predictions, use command *"pred = neigh.predict(data_test)"*.

Step 5 – To evaluate accuracy, use command *"print ('KNeighbors accuracy score:', accuracy_score(target_test, pred))"*.

With the code above, the required modules were imported in the first step. We then developed the object *"neigh"* of type "KNeighborsClassifier" with the volume of neighbors as *"n_neighbors=3"*. In the next step, the *"fit()"* technique was used to train the algorithm on the training data set. Next, the model was tested on the testing data set using *"predict()"* technique. Finally, the accuracy score was obtained, which could be *"KNeighbors accuracy score: 0.814550580998"*, for instance.

Now that our preferred algorithms have been introduced, the model with the highest accuracy score can be easily selected. But wouldn't it be great if we had a way to compare the distinct models' efficiency visually? In Scikit-Learn, we can use the "Yellowbrick library", which offers techniques for depicting various scoring techniques visually.

Chapter 5: Neural Network Training with TensorFlow

TensorFlow can be defined as a Machine Learning platform providing end-to-end service with a variety of free and open sources. It has a system of multilayered nodes that allow for quick building, training, and deployment of artificial neural networks with large data sets. It is touted as a "simple and flexible architecture to take new ideas from concept to code to state-of-the-art models and publication at a rapid pace". For example, Google uses TensorFlow libraries in their image recognition and speech recognition tools and technologies.

Higher-level APIs such as "tf.estimator" can be used for specifying predefined architectures, such as "linear regressors" or "neural networks". The picture below shows the existing hierarchy of the TensorFlow tool kit:

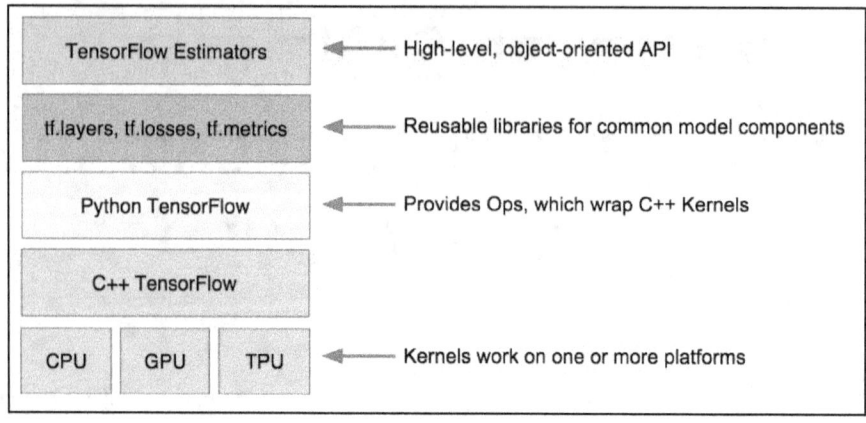

The picture shown below provides the purposes of the different layers:

Toolkit(s)	Description
Estimator (tf.estimator)	High-level, OOP API.
tf.layers/tf.losses/tf.metrics	Libraries for common model components.
TensorFlow	Lower-level APIs

The two fundamental components of TensorFlow are:

1. A "graph protocol buffer"

2. A "runtime" that can execute the graph

The two components mentioned above are similar to the "Python" code and the "Python interpreter". Just as "Python interpreter" can run Python code on several hardware systems, TensorFlow can be operated on various hardware systems, like CPU, GPU, and TPU.

To make a decision regarding which API(s) should be used, you must consider the API offering the highest abstraction level to solve the target problem. Easier to use, but (by design) less flexible, are the greater abstract levels. It is recommended to first begin with the highest-level API and make everything work. If for certain unique modeling issues you need extra flexibility, move one level down. Notice that each level is constructed on the lower-level APIs. It should thus be quite simple to decrease the hierarchy.

For the development of majority of Machine Learning models, we will use "tf.estimator" API, which significantly lowers the number of code lines needed for development. Also, "tf.estimator" is compatible with Scikit-Learn API.

NEURAL NETWORK

The programming of computers needs a human programmer. Many lines of code are used by humans to instruct a computer to provide solutions to our problems. However, the computer can attempt to fix the issue itself through machine learning and neural networks. A neural network is "a function that learns the expected output for a given input from training datasets". For instance, you can train the neural network with many sample bear pictures to construct a neural network that recognizes pictures of a bear. The resulting network operates as functionality to generate the "bear" label as output for the bear picture input. Another more convenient example would be: training the neural network using multiple user activity logs from gaming servers and generate an output stating which users are very likely to convert to paying customer.

Unlike the "Artificial Neural Network" (explained in detail in Chapter 2 of this book), the "Neural Network" features only a single neuron, also called as "perceptron". It is a straightforward and fundamental mechanism which can be implemented with basic math. The primary distinction between traditional programming and a neural network is that computers running on neural network learn from the provided training data set to determine the parameters (weights and prejudice) on their own, without needing any human assistance. Algorithms like "back propagation" and "gradient descent" may be used to train the parameters. It can be stated that the computer tries to increase or decrease every parameter a bit, in the hope that the optimal combination of Parameters can be found, to minimize the error compared with training data set.

FUNDAMENTALS OF NEURAL NETWORK

- Neural networks need clear and informative big data to be trained. You can think of Neural networks as a toddler. They start by observing how their parents are walking. Then they attempt to walk on their own, and the kid learns how to accomplish future tasks with every step. Similarly, the Neural network may fail a few times, but it learns how to generate desired predictions after a few failing attempts.

- For complicated issues such as image processing, it is advisable to use Neural Networks. Neural networks belong to a group of algorithms called "representation learning algorithms". These algorithms are capable of simplifying complicated issues by generating simple (or "representative") form, which tends to be more difficult for conventional (non-representation) algorithms.

- To determine what type of neural network model is suitable for solving the issue at hand, let the data dictate how you fix the issue. For instance, "recurring neural networks" are more appropriate if the issue pertains to sequence generation. While it might be better for you to use "convolutional neural networks" to solve an image-related issue.

- In order to run a deep neural network model, hardware specifications are vital. Neural networks have been around for a long time now, but they are recently experiencing an upsurge primarily credited to the fact that computer resources today are better and more effective. If you want to address a real-life problem using neural network, it is wise to purchase high-end hardware.

TRAINING A NEURAL NETWORK USING TENSORFLOW

In this exercise we will develop a model of neural networks for classifying clothing images such as sneakers and shirts, using TensorFlow library.

I – Import the dataset

For this example, we will be using "Fashion MNIST" data set with 60,000 pictures representing 10 different categories. The low-resolution pictures (28 to 28 pixels) indicate individual clothing items. For the classic MNIST dataset, "Fashion MNIST" is intended as a drop-in replacement. The "MNIST" data set includes pictures of handwritten numbers (0, 1, and so on.) in the same format as the clothing items used in this example. To train the network, we will use 60,000 pictures and 10,000 pictures will be used to assess the accuracy with which the network has learned how to classify pictures.

The "Fashion MNIST" data set is accessible directly from TensorFlow, using the import command as below:

"fashion_mnist = keras.datasets.fashion_mnist (train_images, train_labels), (test_images, test_labels) = fashion_mnist.load_data()"

After the dataset has been loaded system will return 4 different "NumPy arrays" including:

- The *"train_images"* and *"train_labels"* arrays, which serve as the "training dataset" for the model.

- The *"test_images"* and *"test_labels"* arrays, which serve as the "testing dataset" that the model can be tested against.'

Now we need to create labels for an array of integers (0 to 9), corresponding to each category/class of the clothing picture in the data set, using the command below which will look like the table represented in the picture below. This will be useful in generating predictions using our model.

"class_names = ['T-shirt/top', 'Trouser', 'Pullover', 'Dress', 'Coat', 'Sandal', 'Shirt', 'Sneaker', 'Bag', 'Ankle boot']"

Label	Class
0	T-shirt/top
1	Trouser
2	Pullover
3	Dress
4	Coat
5	Sandal
6	Shirt
7	Sneaker
8	Bag
9	Ankle boot

II – Data Exploration

To get some sense of the data set, it can be explored using the commands listed below:

To view the total number of images in the "training data set" and the size of each image – *"train_images.shape"*, which will produce the output displayed as "(60000, 28, 28)" stating we have 60,000 pictures of 28 to 28-pixel size.

To view the total number of labels in the "training dataset" – *"len(train_labels)"*, which will produce the output displayed as "60000" stating we have 60,000 labels in the training data set.

To view the data type of each label used in the "training dataset"– *"train_labels"*, which will produce the output displayed as *"array([9, 0, 0, ..., 3, 0, 5], dtype=uint8)"* stating each label is an integer between 0 and 9.

To view the total number of images in the "testing dataset" and the size of each image – *"test_images.shape"*, which will produce the output displayed as "(10000, 28, 28)" stating we have 10,000 pictures of 28 to 28-pixel size in the testing data set.

To view the total number of labels in the "testing dataset" – *"len(test_labels)"*, which will produce the output displayed as "10000" stating we have 10,000 labels in the testing data set.

III – Data Pre-processing

To make the data suitable for training the model, it needs to be pre-processed. It is essential to pre-process the data sets to be used for training and testing in the same manner.

For instance, you notice the first picture in the training data set has the pixel values between 0 and 255, using the commands below:

"plt.figure()"

"plt.imshow(train_images[0])"

"plt.colorbar()"

"plt.grid(False)"

"plt.show()"

These pixel values need to be scaled to fall between 0 to 1, prior to being used as input for the Neural Network model. Therefore, the values need to be divided by 255, for both the data subsets, using commands below:

"train_images = train_images / 255.0"

"test_images = test_images / 255.0"

The final pre-processing step here would be to make sure that the data is is desired format prior to building the Neural Network by viewing the first 20 pictures from the training dataset and displaying the "class name" under each picture, using commands below:

"plt.figure(figsize=(10,10))"

"for i in range(20):

  plt.subplot(5,5,i+1)

  plt.xticks([])

  plt.yticks([])

  plt.grid(False)

  plt.imshow(train_images[i], cmap=plt.cm.binary)

  plt.xlabel(class_names[train_labels[i]])"

"plt.show()"

IV – Building the Neural Network Model

To build up the "Neural Network", the constituting layers of the model first need to be configured and only then the model can be compiled.

Configuring the layers

The "layers" are the fundamental construction block of a neural network. These "layers" take out information from the data entered generating representations that tend to be extremely valuable addressing the problem.

Majority of "deep learning" involves stacking and linking fundamental layers together. The parameters that are learned during practice are available in most of the layers, like "tf.keras.layers.Dense". To configure the required layers, use command below:

"*model = keras.Sequential([*
  keras.layers.Flatten(input_shape=(28, 28)),
  keras.layers.Dense(128, activation=tf.nn.relu),
  keras.layers.Dense(10, activation=tf.nn.softmax)
])"

The "*tf.keras.layers.Flatten*" is the first layer in this network, which turns the picture format from a 2-dimesnional array of 28x28 size to a 1-dimension array with "28x28 = 784" pixels. Consider this layer as unchained rows of pixels in the picture that arranged these pictures but without any learning parameters and capable of only altering the data.

The network comprises of a couple of *"tf.keras.layers.Dense"* layers after pixels are flattened. These are neural layers that are fully or densely connected. There are 128 nodes or neurons in the first Dense layer. The succeeding and final layer is a 10-node layer of *"Softmax"*, which generated an array of ten different probability scores amounting to "1". Every single node includes a probability score indicating that one of the ten classes is likely to contain the existing picture.

Compiling the model

Before being able to train the model, some final tweaks are needed to be made in the model compilation step, such as:

Loss function— This provides a measure of the model's accuracy during training. This feature should be minimized so that the model is "directed" in the correct direction.

Optimizer —These are the updates made to the model based on the data it can view as well as its "loss function".

Metrics — Used for monitoring the training and testing procedures. For example, the code below utilizes accuracy, measured by computing the fraction of the pictures that were classified accurately.

"model.compile(optimizer='adam',
    loss='sparse_categorical_crossentropy',
    metrics=['accuracy'])"

V – Training the Model

The steps listed below are used to train the "Neural Network Model":

- Feed the training data to the model, using *"train_images"* and *"train_labels"* arrays.

- Allow the network to learn an association of pictures and corresponding labels.

- Generate predictions using the model for a predefined test data set, for example, the *"test_images"* array. Then the predictions must be verified by matching the labels from the *"test_labels"* array.

You can begin to train the network, by utilizing the *"model.fit"* method. To verify the system is a "fit" for the training data, use command *"model.fit(train_images, train_labels, epochs=5)"*.

The epochs are displayed as below, suggesting that the model has reached accuracy of around 0.89 or 89% of the training data:

"Epoch 1/5

60000/60000 [==============================] - 4s 75us/sample - loss: 0.5018 - acc: 0.8241

Epoch 2/5

60000/60000 [==============================] - 4s 71us/sample - loss: 0.3763 - acc: 0.8643

Epoch 3/5

60000/60000 [==============================] - 4s 71us/sample - loss: 0.3382 - acc: 0.8777

Epoch 4/5

60000/60000 [==============================] - 4s 72us/sample - loss: 0.3138 - acc: 0.8846

Epoch 5/5

60000/60000 [==============================] - 4s 72us/sample - loss: 0.2967 - acc: 0.8897

"<tensorflow.python.keras.callbacks.History at 0x7f65fb64b5c0>"

VI – Measuring the accuracy of the Neural Network Model

To test the accuracy of the network, it must be verified against the testing data set using commands below:

"test_loss, test_acc = model.evaluate(test_images, test_labels)"
"print('Test accuracy:', test_acc)"

The output can be obtained as shown below, which suggests that the accuracy of the test result is around 0.86 or 86%, which is slightly less that the accuracy of the training data set. This is a classic example of "overfitting", when the performance or accuracy of the model is lower on new input or testing data than the training data.

"10000/10000 [==============================] - 1s 51us/sample - loss: 0.3653 - acc: 0.8671

Test accuracy: 0.8671"

VII – Generate predictions using the Neural Network Model

Now that our model has been trained sufficiently, we are ready to generate predictions from the model, using command "predictions = model.predict(test_images)".

In the code below, the network has generated a prediction for labels of each picture in the testing data set. The prediction is generated as an array of ten integers with the "confidence" index for each of the ten categories (in referring to the import data stage) corresponding to the test picture.

"predictions[0]"

"array([6.58371528e-06, 1.36480646e-10, 4.17183337e-08, 1.15178166e-10,

8.30939484e-07, 1.49914682e-01, 3.11488043e-06, 4.63472381e-02,

6.10820061e-05, 8.03666413e-01], dtype=float32)"

To view the label with the highest "confidence" index, using command "np.argmax (predictions[0])".

A result "9", will suggest that the model has maximum confidence on the test image belonging to "class_names[9]" or according to our labels table, ankle boot. To verify this prediction, use command "test_labels[0]", which should generate output as "9".

To view the whole set of predictions for the ten classes, use command below:

"*def plot_image(i, predictions_array, true_label, img):*
 predictions_array, true_label, img = predictions_array[i], true_label[i], img[i]
 plt.grid(False)
 plt.xticks([])
 plt.yticks([])

 plt.imshow(img, cmap=plt.cm.binary)

 predicted_label = np.argmax(predictions_array)
 if predicted_label == true_label:
  color = 'blue'
 else:
  color = 'red'

 plt.xlabel('{} {:2.0f}% ({})'.format(class_names[predicted_label],
              100*np.max(predictions_array),
              class_names[true_label]),
              color=color)

def plot_value_array(i, predictions_array, true_label):
 predictions_array, true_label = predictions_array[i], true_label[i]
 plt.grid(False)

plt.xticks([])
plt.yticks([])
thisplot = plt.bar(range(10), predictions_array, color='#777777')
plt.ylim([0, 1])
predicted_label = np.argmax(predictions_array)

thisplot[predicted_label].set_color('red')
thisplot[true_label].set_color('blue')"

Now, for example, you may want to generate a prediction for a specific picture in the testing data set. You can do this using the command below:

"Grab an image from the test dataset
img = test_images[0]
print(img.shape)"

"(28, 28)"

To use the "tf.keras" models to generate this prediction, the picture must be added to a list, since these models have been optimized to generate predictions on a "collection of dataset" at a time. Use the command below to accomplish this:

"# Add the image to a batch where it's the only member.
img = (np.expand_dims(img,0))

print(img.shape)"

"(1, 28, 28)"

Now, to generate the prediction for the picture using "tf.keras" use the command below:

"predictions_single = model.predict(img)
print(predictions_single)"

The predictions generated will resemble the code below:

"[[6.5837266e-06 1.3648087e-10 4.1718483e-08 1.1517859e-10 8.3093937e-07

 1.4991476e-01 3.1148918e-06 4.6347316e-02 6.1082108e-05 8.0366623e-01]]"

To generate a graph or plot for the prediction (as shown in the picture below), use command below:

"plot_value_array(0, predictions_single, test_labels)
plt.xticks(range(10), class_names, rotation=45)
plt.show()"

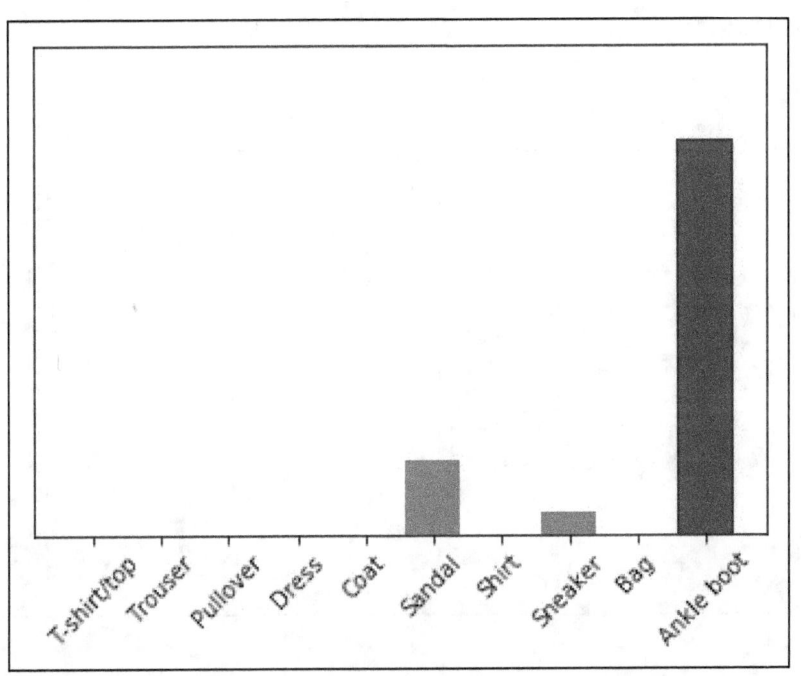

The "model.predict" generated the output as a "list of lists", for every single picture in the testing data set. To generate predictions specifically for the specific image we used earlier, use command below:

*"prediction_result = np.argmax(predictions_single[0])
print(prediction_result)"*

The output or prediction generated should be "9" as we obtained earlier.

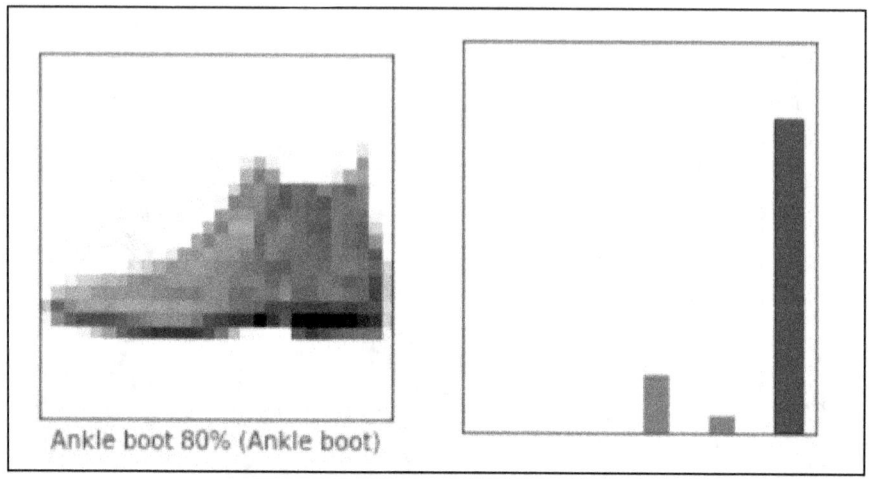

Chapter 6: Data Pre-processing and Creation of Training Data Set

Data Preprocessing is a "data mining technique, which is used to transform raw data into a comprehensible and effective format". Real-world data tend to lack certain behaviors or trends and is almost always incomplete, inconsistent and/or missing attribute values, flooded with errors or outliers. Preprocessing data is a proven way to solve such problems. This raw data or real data from the world can not be readily transmitted through a machine learning model. Therefore, before feeding real-world data to a machine learning model, we need to clean and pre-process it.

OVERVIEW DATA PREPROCESSING

Data Cleaning

Many meaningless and missing sections can be found in the data. "Data cleaning" is performed to manage these inadequacies and constitutes handling data set that is missing values and consists of noisy data.

Missing Data

In this scenario, certain significant information in the data set is missing. It can be dealt with in different respects such as:

Ignoring the tuples: This strategy is appropriate only if the dataset is big and numerous values within a tuple are lacking.

Fill the missing values: This assignment can be done in different ways such as: manually completing the missing values using the mean attribute or the most relevant value.

Noisy Data

"Noisy data" is useless data that machines or the machine learning model is unable to interpret. It can be generated as a result of defective data collection or mistakes in data entry, among others. It can be addressed using the methods below:

Binning Method: This technique operates on sorted data to smoothen it out. The entire data set is split into equivalent size sections and then different techniques are used to finish the job. Each section is fixed individually. To fix the entire data set in ago, all data points in a section can be substituted with its "mean" or most probable values.

Regression: In this case, data can be smoothed out by fitting into a "regression function", which can be either "linear" (with one autonomous variable) or "multiple" (with various autonomous variables).

Clustering: This technique is used to group comparable data points into a cluster. The outliers could be obtained with data points falling outside of the clusters or could not be detected.

Data Transformation

This technique is used to convert the data into a format which is suitable for the data mining method. This includes the following ways:

Normalization: This technique is used to scale data values within a defined range, for example, "-1.0 to 1.0" or "0.0 to 1.0".

Attribute Selection: New data attributes can be generated from an existing data set of characteristics, using this technique to assist in the data mining process.

Discretization: This technique is used to "replace raw values of numerical attributes with interval or conceptual levels".

Generation of the concept of hierarchy: This technique is used to transform lower-level data attributes to higher level in the hierarchical setup. For example, you can convert the attribute "city" to "country."

Data Reduction

"Data mining" is a method used for analysis and extraction of insights from Big data. In such instances, analysis becomes more and more difficult to work with given the enormity of data. We use data reduction method to reduce the volume of data set to an optimal and manageable volume. With this method, the cost of data storage and analysis can be significantly lowered while improving the effectiveness of the data storage. It can be dealt with in different respects such as:

Data Cube Aggregation: This technique is used to apply "aggregation operation" to data to effectively build data cubes.

Selection of attribute subset: This technique is used to ensure that only necessary data attributes are used and the not so relevant attributes can be discarded. To perform attribute selection, the "level of significance" and "p-value of the attribute" can be leveraged. The attribute with "p-value" higher than the "significance level" can be removed to obtain the optimal volume of the data set.

Numerosity Reduction: This technique allows the data model to be stored instead of the whole data set or raw data collected from various input sources, for instance, "Regression Models".

Dimensionality Reduction: This technique uses encoding mechanisms to reduce the volume of the data set. If initial data set can be recovered after reconstruction from compressed data set, this reduction in dimensions of the data set is known as "lossless reduction", otherwise it is referred to as "loss reduction". The two efficient techniques of reducing data set "dimensionality" are: "Wavelet transforms" and "PCA (Principal Component Analysis)".

STEPS OF DATA PRE-PROCESSING

I – Import the data library

There is a wide variety of data libraries available that you can choose to meet your data requirements, such as:

"Pandas": Widely used for data visualization and data manipulation processes.

"NumPy": A basic package to perform scientific computations using Python.

"Matplotlib": A standard Python Library used by data scientists to create 2-D plots and graphs.

"Seaborn": Seaborn is derived from the "Matplotlib" library and an extremely popular visualization library.

For example, you can use the main libraries from "Pandas", "NumPy" and "time"; data visualization libraries from "Matplotlib" and "Seaborn"; and Scikit-Learn libraries for the data preprocessing techniques and algorithms.

To import the libraries mentioned above, use the code below:

For main libraries
*"import pandas as pd
import numpy as np
import time"*

For visualization libraries
*"from matplotlib import pyplot as plt
import seaborn as sns
from mpl_toolkits.mplot3d import Axes3D
plt.style.use('ggplot')"*

For Scikit-Learn libraries

"from sklearn.neighbors import KNeighborsClassifier
from sklearn.model_selection import train_test_split
from sklearn.preprocessing import normalize
from sklearn.metrics import confusion_matrix,accuracy_score,precision_score,recall_score ,f1_score,matthews_corrcoef,classification_report,roc_curve
from sklearn.externals import joblib
from sklearn.preprocessing import StandardScaler
from sklearn.decomposition import PCA"

II – Data exploration

To get some sense of the imported dataset in Pandas, use the code below:

"# Read the data in the CSV file using pandas
df = pd.read_csv('../input/creditcard.csv')
df.head()"

III – Check for missing values

It is essential to comprehend the concept of missing values to be able to effectively manage data. If the researcher does not handle the missing values correctly, they may end up drawing incorrect data inferences. Because of improper handling, the results produced will be different from those with missing values. You can apply any of the techniques below to deal with missing data values in your data set:

1. Ignore the data row

This is generally performed when the "class label" is missing or if a multiple data attribute is missing in the row, assuming the data mining objective is classification. However, if the proportion of such rows with missing class labels is high, you will get bad output.

For instance, database with enrolment data for the student (age, SAT score, address, etc.) containing a column with "Low", "Medium" and "High" to classify their achievement in college. Assuming the objective is to construct a model that predicts the college achievement of a student. Data rows that do not include the achievement column are not helpful to generate predictions regarding the success of the student, so they can be overlooked and deleted before the algorithm is executed.

2. Use a global constant to fill in for missing values

In this technique, an appropriate and new global constant value is selected, such as "unknown," "N / A" or "minus infinity", which is then used to fill all the missing values. This technique is employed when the concerted effort to predict the missing value just doesn't make sense. Let's consider the student enrollment database example again, assume that some students lack information on the 'state of residence' attribute. It doesn't make sense to fill it with some random state instead of using "N/A".

3. Use attribute mean

This technique is used to replace an attribute's missing values with the "mean or median value (if it's discrete)", for a specific attribute in the database. For instance, in a US family income database, if the average income of a family is X, that value can be used to replace missing values in the other family records.

4. **Using attribute mean for all samples belonging to the same class**

This technique is used to restrict the calculations to a particular class in order to obtain a value that is applicable to the row that we are searching for in lies of using the "mean (or medians)" of a particular attribute calculated by searching all the rows of the database. For example, if you have an automobile price database that classifies vehicles, among other things, into "Luxury" and "Low Budget". It is likely more precise to replace the missing price of a "luxury car" with the average price of all luxury vehicles instead of the value obtained after factoring in "low budget cars".

5. **Use data mining algorithm to predict the most probable value**

Data mining algorithms such as "Regression", "inference-based tools using Bayesian formalism", "decision trees", "clustering algorithms (K-Mean\Median, etc.)" can be used to determine the probable value of the data attribute. For instance, "clustering algorithms" could be employed to produce a cluster of rows that are then used to calculate the mean or median of the attribute as indicated earlier in Technique 3. Another instance might be to use a "decision tree" to generate a prediction for the most probable value of the missing attribute by taking into account all other attributes in the data set.

IV – Dealing with Categorical Data values

Categorical attributes can only take on a restricted amount of feasible values, which are generally fixed. For instance, if a dataset is about user-related data, then characteristics such as 'nation', 'gender', 'age group', etc. will constitute the data set. Alternatively, you will find attributes such as 'product type', 'manufacturer', 'vendor', and so on if your data set pertains to a commodity or product.

In the context of the data set, these are all categorical attributes. Typically, these attributes are stored as text values representing different characteristics of the observations. Gender is defined, for instance, as "male" or "female", and product type could be defined as "electronics", "apparel", "food" and so on.

There are three types of categorical data:

- **Nominal** – The types of attributes where categories are only labeled and have no order of succession are called "nominal features". For example, gender could be 'male' (M) or 'female' (F) and have no order of precedence.

- **Ordinal** – The types of attributes where categories are labeled with an order of precedence are called "ordinal features". For example, an economic status feature can contain three categories: "low", "medium" and "high", which have an inherent order associated with them.

- **Continuous** – The types of attributes where categories are numerical variables with infinite values ranging between two defined values are called "continuous features".

CHALLENGES OF CATEGORICAL DATA

- Categorical attributes can contain multiple levels, called "high cardinality" (e.g. states, towns or URLs), where most levels appear in a relatively smaller number of instances.

- Various ML models are algebraic, such as "regression" and "SVM", which require numerical input. Categories have to be changed first to numbers to use these models before the machine learning algorithm can be applied.

- While some machine learning packages or libraries are capable of automatically transforming the categorical data into numeric, depending on the default embedded technique, a variety of machine learning libraries don't support categorical data inputs.

- Categorical data for the computer does not translate the context or background, that people can readily associate with and comprehend. For instance, consider a function called "City" with different city names like "New York", "New Jersey", and "New Delhi". People know that "New York" is strongly linked to "New Jersey" being two neighboring states of America, while "New York" and "New Delhi" are very distinct. On the other hand, for the machine, all three cities just denote three distinct levels of the same "City" function. Without specifying adequate context through data for the model, differentiating between extremely distinct levels will be difficult for it.

ENCODING CATEGORICAL DATA

Machine learning models are built on mathematical equations, so it is easy to comprehend that maintaining the categorical data in equations would cause issues since equations are primarily driven by numbers alone. To cross this hurdle, the categorical features can be encoded to numeric quantities.

The encoding techniques below will be described using example of an "airline carrier" column from a make-believe airline database, for ease of understanding. However, it is possible to extend the same techniques to any desired column.

1. **Replacing the categorical values**

This is a fundamental technique of replacing the categorical data values with required integers. The *"replace()"* function in Pandas, can be used for this technique. Depending on your business requirements, desired numbers can be easily assigned to the categorical values.

2. Encoding Labels

The technique of converting categorical values in a column to a number is called as "label encoding". Numerical labels always range from "0" to "n categories-1". Encoding a group of categories to a certain numerical value and then encoding all other categories to another numerical value can be done using the *"where()"* function in NumPy. For example, one could encode all the "US airline carriers" to value "1" and all other carriers can be given value "0". You can perform similar label encoding using "Scikit-Learn's LabelEncoder".

Label encoding is fairly intuitive and simple and produces satisfactory performance from your learning algorithm. However, the algorithm is at a disadvantage and may misinterpret numerical values. For example, an algorithm may confuse whether the "U.S. airline carrier" (encoded to

6) should be given 6 times more weight "U.S. airline carrier" (encoded to 1).

3. One-Hot encoding

To resolve the misinterpretation issue of the machine learning algorithm generated by the "label encoding" technique, each categorical data value can be transformed into a new column and that new column can be allocated a '1' or '0' (True/False) value, and is called as "one-hot encoding".

Of all the machine learning libraries in the market that offer "one-hot encoding", the easiest one is *"get_dummies()"* technique in "Pandas", which is appropriately titled given the fact that dummy/indicator data variables such as "1" or "0" are created. In its preprocessing module, Scikit-Learn also supports "one-hot encoding" in its pre-processing module via "LabelBinarizer" and "OneHotEncoder" techniques.

While "one-hot encoding" addresses the issue of misinterpreted category weights, it gives rise to another issue. Creation of multiple new columns to solve this category weight problem for numerous categories can lead to a "curse of dimensionality". The logic behind "curse of dimensionality" is that some equations simply stop functioning correctly in high-dimensional spaces.

4. Binary encoding

This method initially encodes the categories as "ordinal", then converts these integers into a binary string, and then divides digits of that binary code into distinct columns. Therefore, the data is encoded in only a few dimensions, unlike the "one-hot encoding" method.

There are several options to implement binary encoding in your machine learning model but the easiest option is to install "category_encoders" library. This can be done using "pip install category_encoders" on cmd.

5. Backward difference encoding

This "backward difference encoding" method falls within the "contrast coding scheme" for categorical attributes. A "K" category or level characteristic typically enters a "regression" as a series of dummy "K-1" variables. This technique works by drawing a comparison between the "mean" of the dependent variable for a level with the "mean" of the dependent variable in the preceding stage. This kind of encoding is widely used for a "nominal" or an "ordinal "variable".

The code structure for this technique is quite similar to any other technique in the "category_encoders" library, except the run command for this technique is "BackwardDifferenceEncoder".

6. Miscellaneous features

You may sometimes deal with categorical columns that indicate the range of values in observation points, for instance, an 'age' column can contain categories such as '0-

20', '20-40', '40-60' etc. While there may be many methods to handle such attributes, the most popular ones are:

A. Dividing the categorical value ranges into two distinct columns, by first creating a dummy data frame with just one feature as "age" and then splitting the column on the delimiter "(-)" into two columns "start" and "end" using "*split()*" and "*lambda()*" functions.

B. Replacing the categorical value ranges with a selected measure like the mean value of the range, using the function "*split_mean()*".

V – Splitting the data set into Training and Testing data subsets

The machine learning algorithms are required to learn from sample data set to be able to generate predictions from the input data set. In general, we divide the data set into a proportion of 70:30 or 80:20, which means that 70% of the data is used as the training subset and 30% of the data is used as the testing subset. However, this split ratio is adjusted according to the form and size of the data set.

It is almost impossible and futile to manually split the data set while making sure the data set is divided randomly. The Scikit-Learn library offers us a tool called the "Model Selection library", to assist with this task. There is a "class" in the Scikit-Learn library called *"train_test_split"*. Using this, we can readily divide the data set into the "training" and "testing" datasets in desired ratios. Some parameters to consider while using this tool are:

- **Test_size** - It helps in determining the size of the data to be divided as the testing data set, like a fraction of the total data set. For example, entering 0.3 as the "test_size" value, the data set will be divided at 30 percent as the test data set. If you specify this parameter, the next parameter may be ignored.

- **Train_size** – This parameter is only specified if the "test_size" has not been specified already. The process works similar to the "test_size" function, except that the percentage of the data set specified is for the "training set".

- **Random_state** - An integer is entered here for the Scikit-Learn class, based on which the "random number generator" will be activated during the data set split. Alternatively, an instance of "RandomState" class can b entered that will then generate random numbers. If you don't enter either of the functions, the default will be activated which leverages the "RandomState" instance used by "np.random".

For example, the data set in the picture below can be split into two subsets: 'X' subset for the "independent features" and 'Y' subset for the "dependent variables" and also happens to be the last column of the data set.

Country	Age	Salary	Purchased
France	44	72000	No
Spain	27	48000	Yes
Germany	30	54000	No
Spain	38	61000	No
Germany	40	nan	Yes
France	35	58000	Yes
Spain	nan	52000	No
France	48	79000	Yes
Germany	50	83000	No
France	37	67000	Yes

Now we can use the code below to split the "x" data set into two subsets: "xTrain" and "xTest" and likewise, split the "y" data set into two subsets "yTrain" and "yTest".

"from sklearn.model_selection import train_test_split
xTrain, xTest, yTrain, yTest = train_test_split(x, y, test_size = 0.2, random_state = 0)"

According to the code above, the test data set size will be 0.2 or 20% of the entire data set and the remaining 80% of the data set will be used for the training data set.

PYTHON TIPS AND TRICKS FOR DEVELOPERS

Python was first implemented in 1989 and is regarded as highly user-friendly and simple to learn programming language for entry-level coders and amateurs. This is ideal for individuals newly interested in programming or coding and requires to comprehend programming fundamentals. This stems from the fact that Python reads almost the same as the English language. Therefore, it requires less time to understand how the language works and focus can be directed in learning the basics of programming.

Python is an interpreted language that supports automatic memory management and object-oriented programming. This extremely intuitive and flexible programming language can be used for coding projects such as machine learning algorithms, web applications, data mining and visualization, game development.

Some of the tips and tricks you can leverage to sharpen up your Python programming skill set are:

In-place swapping of two numbers:
"x, y = 100, 200

print(x, y)

x, y = y, x

print(x, y)"

Resulting Output =

100 200

200 100

Reversing a string:
"a ="machine""

"print("Reverse is", a[::-1])"

Resulting Output =

Reverse is enihcam.

Creating a single string from multiple list elements:
"a = ["machine", "learning", "algorithms"]"

print(" ".join(a))"

Resulting Output =

machine learning algorithms

Stacking of comparison operators:
"n = 10

result = 1 < n < 20

print(result)

result = 1 > n <= 9

print(result)"

Resulting Output =

True

False

Print the file path of the imported modules:
"import os;

import socket;

print(os)

print(socket"

Resulting Output =

"<module 'os' from '/usr/lib/python3.5/os.py'>

<module 'socket' from '/usr/lib/python3.5/socket.py'>"

Use of enums in Python:
"class MyName:

  Geeks, For, Geeks = range(3)

print(MyName.Geeks)

print(MyName.For)

print(MyName.Geeks)"

Resulting Output =

2

1

2

Return multiple values from functions:
"def x():

  return 1, 2, 3, 4

a, b, c, d = x()

print(a, b, c, d)"

Resulting Output =

"1 2 3 4"

Identify the value with highest frequency:
"test = [1, 2, 3, 4, 2, 2, 3, 1, 4, 4, 4]

print(max(set(test), key = test.count))"

Resulting Output =

4

Check the memory usage of an object:
"import sys

x = 1

print(sys.getsizeof(x))"

Resulting Output =

28

Printing a string N times:
"n = 2;

a ="machinelearning";

*print(a * n);"*

Resulting Output =

machinelearningmachinelearningmachinelearning

Identify anagrams:
"from collections import Counter

def is_anagram(str1, str2):

   return Counter(str1) == Counter(str2)

print(is_anagram('geek', 'eegk'))

print(is_anagram('geek', 'peek'))"

Resulting Output =

True

False

Transposing a matrix:
"mat = [[1, 2, 3], [4, 5, 6]]

zip(*mat)"

Resulting Output =

[(1, 4), (2, 5), (3, 6)]

Print a repeated string without using loops:
"print "machine"*3 +' '+"learning"*4"

Resulting Output =

Machinemachinemachine learninglearninglearninglearning

Measure the code execution time:
"import time"

"startTime = time.time()"

" write your code or functions calls"

" write your code or functions calls"

"endTime = time.time()"

"totalTime = endTime – startTime"

"print('Total time required to execute code is=' , totalTime)"

Resulting Output =

Total time

Obtain the difference between two lists:
"list1 = ['Scott', 'Eric', 'Kelly', 'Emma', 'Smith']

list2 = ['Scott', 'Eric', 'Kelly']

set1 = set(list1)

set2 = set(list2)

list3 = list(set1.symmetric_difference(set2))

print(list3)"

Resulting Output =

list3 = ['Emma', 'Smith]

Calculate the memory being used by an object in Python:

"import sys"

"list1 = ['Scott', 'Eric', 'Kelly', 'Emma', 'Smith']"

"print("size of list = ",sys.getsizeof(list1))"

"name = 'pynative.com'"

"print('size of name =' ,sys.getsizeof(name))"

Resulting Output =

('size of list = ', 112)

('size of name = ', 49)

Removing duplicate items from the list:

"listNumbers = [20, 22, 24, 26, 28, 28, 20, 30, 24]"

"print ('Original=' , listNumbers)"

"listNumbers = list(set(listNumbers))"

"print ('After removing duplicate= ' , listNumbers)"

Resulting Output =

"'Original= ', [20, 22, 24, 26, 28, 28, 20, 30, 24]"

"'After removing duplicate= ', [20, 22, 24, 26, 28, 30]"

Find if a list contains identical elements:
"listOne = [20, 20, 20, 20]

print('All element are duplicate in listOne', listOne.count(listOne[0]) == len(listOne))

listTwo = [20, 20, 20, 50]

print('All element are duplicate in listTwo', listTwo.count(listTwo[0]) == len(listTwo))"

Resulting Output =

"'All element are duplicate in listOne', True"

"'All element are duplicate in listTwo', False"

Efficiently compare two unordered lists:
"from collections import Counter

one = [33, 22, 11, 44, 55]

two = [22, 11, 44, 55, 33]

print('is two list are b equal', Counter(one) == Counter(two))"

Resulting Output =

"'is two list are b equal', True"

Check if list contains all unique elements:
"*def isUnique(item):*

tempSet = set()

return not any(i in tempSet or tempSet.add(i) for i in item)

listOne = [123, 345, 456, 23, 567]

print('All List elements are Unique' , isUnique(listOne))

listTwo = [123, 345, 567, 23, 567]

print('All List elements are Unique' , isUnique(listTwo))"

Resulting Output =

"All List elements are Unique True"

"All List elements are Unique False"

Convert Byte into String:
"byteVar = b"pynative""

"str = str(byteVar.decode('utf-8'))"

"print('Byte to string is' , str)"

Resulting Output =

"Byte to string is pynative"

Merge two dictionaries into a single expression:
"currentEmployee = {1: 'Scott', 2: 'Eric', 3:'Kelly'}

formerEmployee = {2: 'Eric', 4: 'Emma'}

def merge_dicts(dictOne, dictTwo):

dictThree = dictOne.copy()

dictThree.update(dictTwo)

return dictThree

print(merge_dicts(currentEmployee, formerEmployee))"

Conclusion

Thank you for making it through to the end of *Python Machine Learning: Discover the essentials of machine learning, data analysis, data science, data mining and artificial intelligence using Python Code with Python tricks*, let's hope it was informative and able to provide you with all of the tools you need to achieve your goals whatever they may be.

The next step is to make the best use of your new-found wisdom in today's cutting-edge technologies, primarily machine learning, that have created the "Silicon Valley" powerhouse. Today, machine learning technology has given rise to sophisticated machines, which can study human behavior and activity in order to recognize fundamental patterns of human behavior and exactly predict which products and services consumer may be interested in. Under the side of their business model, businesses with an eye for the future gradually become technology firms with systems built upon machine learning algorithms. Consider some of the most innovative tech gadgets of this era such as "Amazon Alexa", "Apple's Siri" and "Google Home", what they all have in common is their underlying machine learning capabilities. Now that you have finished reading this book and mastered the use of Scikit-Learn and TensorFlow libraries, you are all set to start developing your own Python machine learning model using all the open sources readily available and explicitly mentioned in this book for that purpose.

If you found this useful you could also like:

MACHINE LEARNING MATHEMATICS

Study Deep Learning Through Data Science. How to Build Artificial Intelligence Through Concepts of Statistics, Algorithms, Analysis, and Data Mining

By Samuel Hack

I would like to thank you for reading this book and if you enjoyed it I would appreciate your review on Amazon!

www.ingramcontent.com/pod-product-compliance
Lightning Source LLC
Chambersburg PA
CBHW070627220526
45466CB00001B/116